A CULTURAL HISTORY OF MATHEMATICS

VOLUME 5

A Cultural History of Mathematics

General Editors: David E. Rowe and Joseph W. Dauben

Volume 1

A Cultural History of Mathematics in Antiquity

Edited by Michael N. Fried

Volume 2

A Cultural History of Mathematics in the Medieval Age

Edited by Joseph W. Dauben, Clemency Montelle, and Kim Plofker

Volume 3

A Cultural History of Mathematics in the Early Modern Age

Edited by Jeanne Peiffer and Volker R. Remmert

Volume 4

A Cultural History of Mathematics in the Eighteenth Century

Edited by Maarten Bullynck

Volume 5

A Cultural History of Mathematics in the Nineteenth Century

Edited by Tom Archibald and David E. Rowe

Volume 6

A Cultural History of Mathematics in the Modern Age

Edited by Tom Archibald and David E. Rowe

A CULTURAL HISTORY OF MATHEMATICS

IN THE NINETEENTH CENTURY

VOLUME 5

Edited by Tom Archibald and David E. Rowe

BLOOMSBURY ACADEMIC
LONDON • NEW YORK • OXFORD • NEW DELHI • SYDNEY

BLOOMSBURY ACADEMIC
Bloomsbury Publishing Plc
50 Bedford Square, 1385 Broadway, New York, NY 10018, USA
29 Earlsfort Terrace, Dublin 2, Ireland London, WC1B 3DP, UK

BLOOMSBURY, BLOOMSBURY ACADEMIC and the Diana logo are trademarks of
Bloomsbury Publishing Plc

First published in Great Britain 2024

Series design by Raven Design
Cover image: The Dame, from Four and Twenty Toilers. (Published 1900) © Historica Graphica Collection / Heritage Images / Getty Images.

A catalogue record for this book is available from the British Library.

A catalog record for this book is available from the Library of Congress.

ISBN: HB: 978-1-3500-6306-8
ePUB: 978-1-3502-9492-9
ePDF: 978-1-3502-9491-2
Set: 978-1-3500-6312-9

Series: The Cultural Histories Series

Typeset by RefineCatch Limited, Bungay, Suffolk
Printed and bound in Great Britain

CONTENTS

ILLUSTRATIONS

FIGURES

Introduction

Chapter 1

Chapter 6

Chapter 7

TABLE

Chapter 5

CONTRIBUTORS

Tom Archibald took his PhD in History and Philosophy of Science at the University of Toronto, Canada, and is currently Professor of Mathematics at Simon Fraser University, Burnaby, Canada. His research concerns the history of mathematical analysis and its applications, and the relationship between mathematics and society more broadly. Until recently, he was co-editor-in-chief of *Historia Mathematica* and a member of the Executive Committee of the International Commission on the History of Mathematics. He now serves on the editorial board of the *Revue d'histoire des mathématiques*. Tom co-authored with Della Dumbaugh and Deborah Kent, "A Mobilized Community: Mathematicians in the United States during the First World War," in David Aubin and Catherine Goldstein eds., *The War of Guns and Mathematics* (2014); and with Rossana Tazzioli, "Integral Equations between Theory and Practice: The Cases of Italy and France to 1920" *Archive for History of Exact Sciences* (2014).

June Barrow-Green is Professor of History of Mathematics at the Open University, UK. Her research focuses on the history of nineteenth- and twentieth-century mathematics, and she has a special interest in the history of the gender gap in mathematics. Among her recent publications is a study of the geometrical surface models of Olaus Henrici (1840–1918). She is chair of the International Commission on the History of Mathematics, and a winner of the Royal Society Wilkins-Bernal-Medawar Medal (2021) and the Chandler Davis Prize for Expository Excellence (2014). "'Knowledge Gained by Experience': *Olaus Henrici*—Engineer, Geometer and Maker of Mathematical Models," *Historia Mathematica* (2021).

Jeremy Gray is Emeritus Professor at the Open University and an Honorary Professor at the University of Warwick, UK. He is interested in the history of mathematics in the nineteenth and early twentieth centuries, its philosophical implications, and modernism in mathematics. In 2009, he was awarded the Albert Leon Whiteman Memorial Prize of the American Mathematical Society. His most recent book is *Change and Variations* (Springer 2021) on the history of differential equations.

Adrian Rice is the Dorothy and Muscoe Garnett Professor of Mathematics at Randolph-Macon College, Virginia, USA, where his research focuses on nineteenth- and twentieth-century mathematics, particularly with regard to developments in Victorian Britain. Recent interests include nineteenth-century mathematics education, the history of elliptic functions, and the mathematics of Ada Lovelace. He is a four-time recipient of awards for outstanding expository writing from the Mathematical Association of America. Adrian co-authored with Christopher Hollings and Ursula Martin, *Ada Lovelace, The Making of a Computer Scientist* (2018).

David E. Rowe is Professor Emeritus of History of Mathematics and Natural Sciences at Mainz University, Germany. His principal research interests concern developments in Western mathematics and physics during the period 1800 to 1950, with special focus on Göttingen University, Germany. Since 2011, he and Klaus Volkert have co-edited the series *Mathematik im Kontext*. He has written or co-edited sixteen books, most recently *Emmy Noether: Mathematician Extraordinaire* (2021).

Tony Royle is Associate Lecturer in Mathematics at the Open University, UK, and a former military and civilian pilot. His main research interest is the part played by mathematicians, scientists, and engineers in the development of aeronautics and aeronautical engineering during the early part of the twentieth century, as described in his recent book, *The Flying Mathematicians of World War I* (2020).

Anja Sattelmacher is a historian of science and media. She took her doctorate in history at the Humboldt-University Berlin, Germany, under Anke te Heesen, who inspired her dissertation on mathematical models. This study was published as *Anschauen, Anfassen, Auffassen. Eine Wissensgeschichte mathematischer Modelle* (2021). She is presently teaching at the Department of Musicology and Media at Humboldt-University, while working on a project dealing with the history of research films.

Ivahn Smadja is Professor of History and Philosophy of Science at the University of Nantes, France. He trained at the École normale supérieure in Paris and the University of Paris Sorbonne. His main interest lies in the history and philosophy of mathematics. His current research project, based on archival work at the Paris Académie des sciences, focuses on the historiography of ancient

(Western and non-Western) mathematics in the nineteenth century. Ivahn's publications include, "On Two Conjectures that Shaped the Historiography of Indeterminate Analysis: Strachey and Chasles on Sanskrit Sources," *Historia Mathematica* (2016); and jointly with Nicola Michel, "Mathematics in the Archives: Deconstructive Historiography and the Shaping of Modern Geometry (1837–1852)," *British Journal for the History of Science* (2021).

Scott A. Walter is Professor of Epistemology and History of Science in the Faculty of Science and Technology in Nantes, France. His lectures and research focus on the technical history and philosophy of contemporary mathematics and physics. Since 2002, he has curated Henri Poincaré's writings for the website Henri Poincaré Papers. He published an essay review of Jeremy Gray's biography of Poincaré (2013) in *Historia Mathematica* (2017).

SERIES PREFACE

Culture, history, and mathematics are words that seldom appear together. Yet, mathematical activities have played a pervasive role in diverse cultures throughout human history, as described in the present six volumes. These illustrate how individuals and societies have engaged with various manifestations of mathematical culture from antiquity to the present. Based on recent and ongoing scholarship, they provide a global overview of a highly diverse range of developments, though certainly nothing like a comprehensive picture. Editors and authors had to make difficult choices about which topics and cultures to include, and in doing so, they have sought to strike a balance between characteristic features of a period and particular individuals and events that had a deep influence or longer lasting impact.

Each volume addresses seven general themes in as many chapters. The first three chapters —Everyday Numeracy, Practice and Profession, and Inventing Mathematics—highlight the roles of those who used, taught, or created mathematics as it evolved over time. These chapters explore mathematics per se, whereas the three that follow—Mathematics and Worldviews, Describing and Understanding the World, Mathematics and Technological Change—consider the larger impact of mathematics on other dimensions of human culture. The final chapter—Representing Mathematics—is devoted to a broader understanding of mathematical ideas, activities, or artifacts in different cultures throughout the world. The periods covered are:

Volume 1: A Cultural History of Mathematics in Antiquity
(3000 BCE–500 CE)
Volume 2: A Cultural History of Mathematics in the Medieval Age
(500–1400)

Volume 3: A Cultural History of Mathematics in the Early Modern Age (1400–1687)
Volume 4: A Cultural History of Mathematics in the Eighteenth Century 1687–1800)
Volume 5: A Cultural History of Mathematics in the Nineteenth Century (1800–1914)
Volume 6: A Cultural History of Mathematics in the Modern Age (1914–2020)

General Editors: David E. Rowe and Joseph W. Dauben

Introduction

TOM ARCHIBALD AND DAVID E. ROWE

Many features of our modern world first became recognizable during the "long" nineteenth century from 1800 to 1914. The historian Eric Hobsbawm divided this period in three, corresponding to an age of revolution, an age of capital, and an age of empire, although revolution, capitalism, and imperialism were all present throughout. In fact, the entire era defines a period in which today's large-scale political divisions began to take shape and during which familiar forms of government and economic activity emerged and matured. Already unified by trade, the world became even more interconnected, as steam replaced sail, railroads reached across continents, and as communication came to include telegraphy and wireless. Even at the beginning of the century, young Friedrich Wilhelm Bessel, who later became a leading astronomer, taught himself advanced mathematics precisely because he wanted to improve his performance as a dispatcher for a large shipping firm in Bremen, Germany, with worldwide reach.

Indeed, mathematics participated in these historical developments in many important ways, as discussed in the chapters of this volume. It served, for example, as a fundamental tool of trade and business (via arithmetic and accounting). Mathematical methods played a major role in political and social organization (for example, through mapmaking and statistical recording), and in waging war (with technological advances in ballistics). Mathematical innovations on a more rarified level were crucial for describing and understanding, but also manipulating the natural world, via modern theoretical physics. At the same time, new mathematical ideas increasingly shaped an autonomous realm of scientific and cultural endeavors, in which the invention of such things as non-Euclidean geometry and mathematical logic led to a reappraisal of the relationship between the human mind and the knowable universe. Thus, the emergence of the modern world took place alongside the development of a modernized mathematics, which was itself part of that larger new world.

The century began explosively with the Napoleonic Wars (1800–15), a series of attempts led by Napoleon Bonaparte to remake Europe and incorporate it into the French Empire. Emerging from the revolution that ended the Bourbon monarchy in 1792, Napoleon began his political career as a general, before becoming first consul and then emperor in 1804. His hold on power ended in 1815, but afterward the map of Central Europe had to be repeatedly redrawn. Germany, a collection of hundreds of mainly small states in 1800, emerged from this period as a loose association of a dozen or so regional and city states, soon to form a customs union, before unifying in 1871. Italy, too, reverted from French rule into a patchwork of monarchies and dukedoms, followed by the *Risorgimento* that culminated with Italian unification around the same time. Britain, victorious, increased its dominance in international trade and expanded its empire at French expense. The French also contributed to the growth of Britain's former colony, the United States, by the sale of the Louisiana territory in 1803. The wars in Europe and their immediate aftermath also provided opportunity for independence movements in Argentina and Mexico (wrested from Spain in 1818 and 1821, respectively), in Brazil (a Portuguese colony until 1822), and a great deal of northern South America, liberated under the leadership of Simón Bolívar.

This initial redrawing of the world map reflected a halting and uneven transition from the kind of aristocratic rule destroyed by the French Revolution of 1789–99 to governments based on constitutional principles and subject to the rule of law. The British Reform Act of 1832, engineered in part by the grandfather of philosopher-mathematician Bertrand Russell, not only extended constitutional controls on the monarchy but also expanded support for a more democratic society by extending the reach of elementary public education. Elsewhere, however, liberal reform movements met with defeat. When liberals attempted to install a modern constitution in the German state of Hanover, its Grand Duke annulled it, leading to a famous protest by seven professors at Göttingen University, an action that cost them their positions. Up until the revolutions of 1848, political activities of students across the German-speaking states came under heavy surveillance.

The Franco-Prussian War of 1870–71 led to the unification of Germany. Italy also went through a complex series of events before forming a unified nation under a single monarch, while leaving the Catholic Church sidelined as a territorial power. The result was a Europe with fewer, but stronger nations, leading to intense competition in varied political, economic, and cultural spheres. Within the civil arena, these rivalries gained peaceful outlets through world's fairs (e.g., London 1851 and Paris 1889), and the Olympic Games (beginning in 1896). As political liberalism waned, the German universities gained prestige by adopting an apolitical form of academic freedom in both teaching as well as learning, driven by the notion that professors and students were scholars engaged in original research. The centralized French model for institutions of higher education was entirely different, although throughout continental Europe the churches no longer exerted their former secular influence. By the end of the period, the transition to modern research-centered universities throughout Europe and beyond its shores was well underway.

Already during the first decades of the century, European technological dominance, notably in weaponry, went hand-in-hand with demands for trade access and even led to political control in several parts of the world. In South Asia, the British East India Company exerted control and, with Crown support, effectively created the Indian state, which came under direct British rule in 1858 following revolts. Mathematics—imported from Europe as a tool of imperial control, notably in mapmaking and census statistics—was taught to limited groups following British models. In East Asia, the Japanese state, reconfigured in 1868 as the Meiji Empire, chose to Europeanize. Traditional Japanese mathematics (Wasan) soon fell out of favor, replaced by Western approaches, as near the end of the century a new generation of leaders began studying in Germany. China resisted Western incursion, a losing battle by stages, leading to the collapse of the empire shortly after the turn of the century and the creation of the Republic of China in 1911. Chinese scholars began to seek European, Japanese, and US education even before the First World War, partly financed by reparation payments from the imperial and then national governments.

The French Revolution had embraced the notion of human rights for a free citizenry, and this ideal spread across Europe and its dependencies in a variety of ways. Here, as emblematic, should be mentioned the abolition of slavery in Britain (1806), the US (in 1864), and Brazil (in 1888); and the reform of land tenure to eliminate serfdom (accomplished earlier in Western Europe and ending when Czar Alexander II emancipated the serfs in Russia in 1861). At the same time, colonialist efforts diversified, with an 1882 meeting of great powers in Berlin essentially dividing much of the globe, in particular the continent of Africa, into possessions of the European states. The notion that all people should be citizens of a nation state gained general acceptance through the Treaty of Versailles at the end of the First World War. This principle of national self-determination, however, did not entail any universal conception of widespread human rights, nor of rights for particular groups, notably women.

Nevertheless, women in Europe and the US eventually gained suffrage. Well before, however, a few had even attained visibility on the mathematical stage, as in other areas of creative endeavor. Sophie Germain attracted curiosity for successfully pursuing higher mathematics as an autodidact in Napoleonic Paris. Similarly, Ada Lovelace, daughter of the poet Lord Byron, developed an interest in mathematics and worked with Charles Babbage on various projects, including his "analytical engine" (Hollings, Martin, and Rice 2018). Decades later, the Russian Sofia Kovalevskaya managed to study mathematics under Karl Weierstrass in Berlin; afterward she obtained a university professorial position in Stockholm, though her death at age 41 ended her promising career as a creative mathematician. By the first decade of the twentieth century, select colleges and universities in Western Europe and the US offered women the chance to pursue higher learning, even if such studies did not always lead to degrees or professional opportunities.

Over the course of the century, as creative mathematics developed enormously in many nations and extended its reach into different fields of science and technology, the field began to function as an international entity. At the beginning of the century, the true lingua franca of advanced mathematics was French, though scholars still read Latin and used it in some contexts. By century's end, journals published mathematics in most of the European languages but also in Japanese. Germany, with a flexible system of higher education that welcomed foreign students, proved more attractive to foreigners than France, where one rarely encountered professors except in their lecture halls. In this way, not least because of charismatic and entrepreneurial figures like Felix Klein and David Hilbert, the German model for higher education came to have particular resonance internationally, notably in the US. In this same period, international contacts led to intense scientific competition and rivalries spread to other parts of the world. In mathematics, local and national meetings in science and mathematics proliferated throughout the century, with the first International Congress of Mathematicians taking place in 1897.

The unparalleled destruction, above all of human life, associated with the Great War—the first true *world* war—led again to a redrawing of the map and a redistribution of colonies and regions of influence. The Austro-Hungarian and Ottoman empires dissolved and Russia became the Soviet Union following the 1917 revolution. Soviet Marxism would vie with Italian and German Fascism as the two new political ideologies that would prepare the stage for the next great military conflict. The vast scale of the First World War already brought science and technology into sharp focus as constitutive parts of military power. Their importance was evident not only through industry and communication—developments that took place over the whole century—but especially during wartime when mathematics played a special role, as in ballistics and wireless technology. This twentieth-century thread of developments took its start as our period closes.

This Cultural History of Mathematics approaches the subject in Chapters 1 to 3 by moving upward—from everyday usages to cutting-edge research—and then outward in Chapters 4 through 7—to explore manifestations of mathematics in various other realms of human culture. These seven chapters address some of the ways in which mathematics interfaces with other societal and cultural spheres. Their aim is to present an accessible portrait and analysis of what is often a recondite field by showing how mathematics makes contact with other perhaps more familiar aspects of life. The authors of the individual chapters have thus, each in their own way, offered a portrayal of mathematics as a culturally embedded human endeavor.

MATHEMATICS AND THE RISE OF PUBLIC EDUCATION

The increased diversity and effectiveness of mathematics in describing and understanding the world, and indeed the expansion of creative mathematics, rested on an increased awareness and use of basic mathematics, above all arithmetic. At the beginning of the century, in countries where trade, commerce and industry flourished, individuals who could manage "figures" had improved chances for employment, and many jobs that traditionally might not have needed calculating skills did so increasingly. In the past, numeracy was largely in the hands of specialists, especially those who wrote books designed for self-instruction, whereas the nineteenth century witnessed a far broader spreading of such knowledge. Expanding economic opportunities brought new possibilities for paid work and upward social mobility, creating more demand among ambitious (or desperate) members of the numerate population.

Beyond harnessing mathematical expertise for the labor market, however, there was a strong impulse to improve the lot of humanity through education. Although this movement took on different forms in various countries, the trend itself was nearly universal, as during the course of the century compulsory public education became the norm in more industrialized societies. In many countries concerns about improvements for the poor led to a drive toward a broadening of educational access for children, especially boys at first. Literacy lay at the heart of this, as did, in some places, a move to civilize the poor by training them to wash, carry handkerchiefs, and know how to act politely to their betters. Arithmetic went along with this due to its undeniable utility.

Over the course of the century, in Western Europe as well as in the US, schooling became compulsory for all, both male and female, up to a certain age, varying with the country. The connection with pre-existing religious schooling varied with the nation, remaining close in Germany and Italy, while separated by law in France. These same countries emphasized schooling in their colonies, at least with the end in mind of providing appropriately skilled labor from the local population, sometimes (as in India) piggy backing on elaborate local education systems. In Chapter 1 of this volume, Adrian Rice focusses primarily on the cases of the UK and the US, though the aims pursued there were quite similar elsewhere, even if the means and organization varied widely from place to place.

Arithmetic was important in trade, of course, and this meant exchanging money. Rice vividly calls to mind the kinds of complications that arose due to local, regional, and national monetary systems. Different systems of non-decimal based weights and measures could easily lead to thorny arithmetical headaches: quick now, how much is a C-weight of sugar at 4d. 3 farthings per pound? The astonishing variety of systems, and the necessity of combining cash and barter exchanges, made such trade-related problems universal. Systems of taxation also frequently depended on such measurements, and hence complicated calculations. The everyday numeracy of the marketplace thus survived, and those who sold their wares directly to the public were often extremely adept at counting with their fingers. Digitizing preceded digitalization, but in fact, it took a very long time for the advantages of base ten—the Indo-Arabic numeral system, which eventually wound its way into Europe—to be fully appreciated.

The system of decimal (base 10) measurements, today referred to as the metric system, was a product of the social reforms of the French Revolution. In 1790, basing their ideas on a more restricted proposal of the Academy from before the revolution, the leaders established a commission of five leading scientists charged with the task of proposing a new system. Pierre Simon de Laplace and Joseph-Louis Lagrange, both famed for their work in celestial mechanics, were among that group. Three years later, on August 1, 1793, the commission adopted the *mètre* as a new unit of length, defined as one ten-millionth of the distance between the North Pole and the equator on a longitude through Paris. By 1795, the revolutionary government passed into law a system of measurements based on the *mètre*. For example, this law introduced the *gramme* as the absolute weight of a cubic centimeter of pure water at the melting temperature of ice.

These innovations had a lasting influence, despite some variations in the regulations, whereas the new system for time measurements proved to be more ephemeral. To observe the brief duration of one second—defined as a certain small fraction of the solar year (one rotation of the earth around the sun)—required a pendulum of appropriate metric length placed at 45 degrees north latitude. The number 10 already enjoyed pride of place in the new Republican calendar, which divided the year into 12 months, each with three ten-day weeks, with five additional days intercalated and a leap day every four years. Compared with the last major calendrical reform—introduced under Pope Gregory XIII in 1582, but only slowly adopted in Protestant and Eastern

Orthodox countries over the course of more than three centuries—the French calendar represented a truly radical break with the past. Still, the desire to rationalize time and enshrine everything in base ten went further; the hour had struck for introducing the decimal clock, which divided the day into ten hours, each hour into 100 decimal minutes, and each decimal minute into 100 decimal seconds. Not surprisingly, this bold idea never caught on at all, and already in 1795, the government abolished this device. The calendar lasted longer and the *mètre* longer still, but the system itself failed to garner public support or capture peoples' imagination. Thus, the old measures survived—for obvious reasons, since distributing systems of weights, for example, would have required enormous resources, let alone teaching people how to use them. Following efforts to determine the basis of the system precisely, for example by measuring the distance from Dunkirk to Barcelona (that is, along a meridian of longitude) by precise surveying methods, the new system took hold.

Despite the difficulties involved, France introduced a law governing metric measures for general use in 1840, though only instituted by the end of the 1850s, at least officially; in reality, local measures persist to the present day. The unification of Germany in 1871 led to the adoption of this same system in 1872, harmonizing dozens of local systems in which different *Meile* contained

FIGURE 0.1. A *mètre* placed in public for general use in Paris. Wikimedia Commons.

different numbers of feet, lengths which themselves varied locally. Italy, largely under French rule in the Napoleonic period, introduced the *mètre* in the Kingdom of Sardinia in 1845. After unification under the former Sardinian king, Vittorio Emmanuele of Savoy, the remainder of the peninsula followed suit. Each of these efforts, and the many others that accompanied the institution of the metric system elsewhere, required a rollout of devices and educational resources. The UK and the US retained their older (related but subtly different) systems, though their schools taught metric units used in science and engineering. The metric system gradually became the standard worldwide over the course of the nineteenth century, a movement in which James Clerk Maxwell, famous for his work in electromagnetic theory, played a central role. Interestingly enough, in 1897, Henri Poincaré spearheaded an ill-fated proposal to standardize decimal units of circumference and time in 1897. It won support, just not enough to overcome the entrenched opposition.

Weights and measures, then, were an important motivating theme for mathematics education at the elementary level, and the need for such knowledge helped promote higher standards and national requirements for such education. Decimal measures had the advantage of ready comprehension by teachers, who were often working near the thresholds of their grasp. Indeed, they rarely imparted arithmetic skills by way of understanding but rather by rule and rote learning. These austere and authoritarian methods hardly produced warm feelings among a teacher's charges, though they surely helped instill discipline in the classroom. Lewis Carroll (the pseudonym of C. L. Dodgson, a mathematics professor at Oxford) hinted at the drudgery that accompanied this type of learning experience in *Through the Looking-Glass* (1871), where he wrote: "Reeling and Writhing of course, to begin with," the Mock Turtle replied, "and the different branches of arithmetic—ambition, distraction, uglification, and derision." For carrying out actual calculations, one could use books of tables—"ready reckoners"—or for those who, like surveyors, did many calculations in the field, by using slide rules. The locus of such teaching was the schoolroom, often with students at many levels working at the same time, writing at first on slates with a stylus or, in more affluent places and periods, in paper copybooks. Blackboards eventually became common.

The qualifications of an elementary schoolteacher were limited, though by the middle of the century teachers' colleges began to appear. Despite the usual emphasis on rote learning, earlier trends promoting reflection and experiment toward a more ideal and effective education continued. Early in the century, for example, Johann Heinrich Pestalozzi ran an institute in Yverdon (Switzerland) that gained international renown for its reasoned psychological approach to the education of children (see Chapter 7). It even drew the attention of some Berlin academics, among them J. L. Cauer, who visited during the immediate post-Napoleonic period when seeking faculty for a new high school. A student of

GEOFFROY. — AT SCHOOL.

FIGURE 0.2. A typical schoolroom scene. Photo by Duncan1890 / Getty Images.

Pestalozzi's, Friedrich Fröbel, introduced various educational innovations, including manipulatives (building blocks) as a step toward teaching geometry. He coined the term "kindergarten" to connote learning based on the special needs and abilities of small children. Another product of Pestalozzi training was Jakob Steiner, who later became professor of geometry in Berlin from 1834 to 1863. Such proposals for educational reform were characteristic of the period, and the progressive education movement in the US from the 1880s, eventually clustered around the work of John Dewey, was influential internationally, though not only in teaching mathematics.

Over the course of the long nineteenth century, literacy and numeracy reinforced one another, as the everyday world became increasingly saturated with numbers in all sorts of cultural contexts. Newspapers everywhere tended to report various forms of statistical information, from populations, to numbers of infections, to sports standings. While extensive records of cricket scores may not seem important, their publication attests to an association of numeracy with everyday life. Less trivial were cases in which numbers were collected, not by mathematicians but by those in other walks of life, in order to draw conclusions of general social importance. A famous example comes from public health statistics gathered in London in the mid-nineteenth century by Dr. John Snow, tracing a cholera outbreak in Soho to a well in Broad Street. While

Snow's arguments did not completely convince officialdom—reminiscent of the present COVID-19 outbreak—he was successful in having the pump handle removed. This kind of use of numerical data surfaced in many spheres of public and private life. Indeed, all sorts of policy decisions became increasingly dependent on the ability to understand and work with numbers. Those who were engaged in mathematical practices relied, in turn, on a broad numerate public that included those employed in legislatures, the military, or in trade and industry.

TRANSFORMATIONS IN PRACTICE AND PROFESSION

A cultural history of mathematics, like this one, is predicated on the assumption that mathematical practice, like any other form of human activity, has been deeply affected by social, political, and technological developments, but particularly so over the last two centuries. Even well before the onset of the electronic age and the IT revolution, the dominant modes of communication used by mathematicians underwent a profound transformation, which had major repercussions for mathematical research. This involved not only written forms of communication but also oral exchanges, which became increasingly important over the course of the long nineteenth century, as travel became easier and personal meetings far more frequent.

When seen against the backdrop of the periods covered in Volumes 3 and 4, the larger developments in mathematics during the nineteenth century appear as a natural outgrowth of broader transformations that affected scientific institutions, networks, and discourse. The leading researchers of those earlier days—from Newton and Leibniz to Euler and Lagrange—did their work in an era dominated by scientific academies and societies under the auspices of royal patrons. Much of their published research appeared in the reports issued by these institutions, some of which continued to play an important role well into the nineteenth century. As professional societies proliferated, their members contributed to shaping new communities with their own distinctive cultural identities. Along with these developments came new mathematical practices. In lecture halls and seminar rooms, the spoken word gradually assumed a central place in making as well as spreading mathematical knowledge. At the same time, the first scientific journals specializing in mathematics arose as outlets for disseminating and circulating newly found results. Some even served as hubs for networks of researchers well before the formation of national mathematical societies.

Up until the nineteenth century, those who taught mathematical subjects and those who contributed to mathematics as scientific researchers often found themselves situated in two largely distinct communities. The highly prolific Leonhard Euler was, to be sure, the author of widely read books on algebra and analysis. Nevertheless, he wrote these, so to speak, during his leisure hours at the St. Petersburg and Berlin Academies. Like most of the "inventive" mathematicians of his era, Euler was not a professional educator, even though he offered morning lecture courses in Saint-Petersburg. He taught these on a regular basis in Russian, a condition for his employment at the Academy.

Within the Anglo-American setting, on the other hand, several leading scientists were amateurs, at least in the sense that their positions did not require research. This was a tradition that persisted after 1800. Nathaniel Bowditch, who eventually translated and wrote commentaries on the first four volumes of Laplace's *Traité de mécanique celeste*, was a New England autodidact who mainly worked as an actuary for an insurance company. In his younger years, though, he had been a ship's clerk with considerable expertise in celestial navigation. This led to the publication of Bowditch's *American Practical Navigator* in 1802, a work studied and used by naval officers well into the twentieth century. The US Naval Observatory's Nautical Almanac Office later employed two of the nation's most eminent mathematicians: George William Hill and Simon Newcomb. Hill was a leading expert on lunar theory, famous above all for Hill's differential equation, which was lauded by Henri Poincaré. Up until 1875, only a handful of those who taught mathematics at colleges and universities in the United States were actively engaged in high-level research, whereas on the continent of Europe the merging of teaching and research began far earlier.

As described in our Chapter 2, this dynamic of professionalization traces back to reforms inaugurated during the French Revolution, which elevated the social status of mathematics, the very backbone of the curriculum at the École polytechnique. This newly founded school trained a new class of engineers, many of whom left a deep mark on French scientific culture (Grattan-Guinness 1990). One of the most influential figures at the École polytechnique was Gaspard Monge, a gifted geometer and teacher who enjoyed close relations with Napoleon. Monge invented a new practical approach to drawing that became a standard method taught to architects and engineers: descriptive geometry. The fundamental idea behind this uses orthogonal projections to represent three-dimensional objects by means of coordinated images in three planes (Chapter 7). Although guarded as a military secret, Monge's methods soon spread to other countries, starting with Italy, Spain, and Portugal, where translations of Monge's *Géométrie descriptive* (1799) appeared between1803 and 1812. By mid-century, descriptive geometry was a firmly established subject in the curricula of numerous polytechnical schools throughout Europe and the Americas (Barbin, Menghini, and Volkert 2019).

Another mathematician who played a central role in French history was Lazare Carnot, who studied at the École royale du génie de Mézières where the curriculum included geometry, mechanics, geometrical designing, geography, and hydraulics. He later published pioneering work in projective geometry, a field made famous by another military engineer, Jean-Victor Poncelet, author of *Traité des propriétés projectives des figures* (1822). Carnot's *Géométrie de position* (1803) gave the first treatment of cross-ratios using signed magnitudes, an important advance for this new direction in research. He was even better known for his works on fortification, which linked him with an older and much esteemed French tradition.

Carnot was largely responsible for building the French Revolutionary Army through conscription as well as training its forces to mobilize quickly and attack opposing armies on their flanks rather than head on. As a rival of Maximilien Robespierre, he managed to survive the Reign of Terror and become a member of the Directory in 1795 when he appointed Napoleon Bonaparte to head the Army of Italy. Five years later, Napoleon appointed him Minister of War in 1800, though he served only briefly and refused political office after his former general crowned himself emperor in 1804. Carnot came to his aide, however, during the disastrous Russian invasion and even served as Minister of the Interior during the Hundred Days. After Waterloo, he lived in exile with his younger son, Hippolyte, and died in Prussia in 1823.

Lazare Carnot's elder son, Sadi, became famous for his analysis of heat engines, an early chapter in the history of thermodynamics (see Chapter 5). After returning to France, Hippolyte Carnot entered politics and became a follower of the utopian socialist Saint-Simon (Chapter 4). His son, whom he named after his now deceased older brother, graduated from the École polytechnique before beginning his own political career. Hippolyte lived just long enough to see his son, Sadi Carnot, become president of France in 1887, but fortunately not to witness his assassination in 1894. These illustrious figures illustrate how career trajectories of French mathematicians and scientists often mingled closely with those of engineers and politicians. A somewhat similar pattern developed later in Italy, whereas nothing at all similar transpired in the German states. Prussia, to be sure, created a modern state bureaucracy, but its civil servants were primarily lawyers by training, while the upper echelons of the military remained firmly in the hands of the hereditary aristocracy.

When Berlin University first opened in 1810, Prussia was still under French occupation, having fallen to Napoleon's *grande armée*. Parisian research mathematicians, both pure and applied, dominated these fields and they maintained that position until the 1830s. True, Germany had Carl Friedrich Gauss, but by his own choice, Gauss interacted mainly with fellow astronomers and physicists. A rough count of Europe's most productive research mathematicians would easily reveal that more of them were living within the walls of Paris than outside them. Indeed, the École polytechnique was turning out throngs of them as a mere by-product of its curriculum for training engineers. Yet, fifty years later, the German universities, led by those in Prussia, would emerge as the foremost training ground for aspiring young mathematicians.

The differences between the French and German educational systems were striking. These reflected two very different organizational models for education, which in France was a centralized national system, whereas in Germany the individual states administered their schools, which thus varied a great deal. Moreover, there were major differences in the role that mathematics played in these two societies. Germany had a number of polytechnic schools, but these enjoyed far less status than the universities, which were only open to young men who graduated from a gymnasium. From a young age, these lads studied ancient languages and literature, often taught by very learned scholars who made serious demands of their charges. Within these elite circles, Latin and Greek were practically the *lingua franca* for educated discourse. Mathematics, too, enjoyed high esteem, though largely for purely formal reasons; it represented yet another type of language that one learned in order to discipline the mind.

Many German intellectuals made no secret of their distaste for the French system, which they associated with materialism and utilitarianism. This attitude was especially prevalent in Prussia, where the Protestant work ethic and a strong sense of duty pervaded everyday life. Prussian military virtues mixed with cultural Romanticism to produce a generation of scholars for whom science (*Wissenschaft*) took on quasi-religious meaning as a higher calling. These ideals persisted up until the fall of the monarchy and even into the era of the Weimar Republic. Max Weber's lecture from 1917, "Science as Vocation" (*Wissenschaft als Beruf*), stands out as an example of sober and critical reflections on this German intellectual tradition.

If French *savant* mathematicians bore responsibility for training a new profession of technocrats, German professors mainly taught future Gymnasium teachers, a position that carried considerable social prestige itself. Indeed, a number of Germany's leading mathematicians, including Kummer and Weierstrass, began their careers as Gymnasium teachers, but scores of others who never dreamt of university careers published respectable work in leading academic journals. This neo-humanist approach to scholarship persisted throughout most of the century at the universities, though by the 1890s it met strong resistance from the institutes of technology. Despite its inherent conservativism, the German model nonetheless proved highly conducive for the formation of modern research schools based on free inquiry, as opposed to the kind of "scholasticism" that continued to dominate so-called higher learning in other European countries.

France and Germany thus spawned two competing models for the mathematical sciences, which exerted a strong influence on practice and professionalization in other countries, both within Europe and beyond its borders. International trends and cooperation grew stronger throughout the period 1870 to 1914, as noted in Chapter 2. British institutions, on the other hand, were slower to modernize, as reflected in the problem-solving tradition of the mathematical Tripos at Cambridge University. When young Americans began taking up graduate studies in Europe in the 1880s, they typically chose German universities or the Sorbonne in Paris rather than attending a British school. Those who went on to assume leadership roles in the United States were motivated to develop a distinctly American tradition that could compete with the best schools in Europe. Indeed, mathematical research during the later decades of the period reflected a strong sense of competition between nations with their own professional societies and educational traditions. Thus, in sharp contrast with the preceding era of scientific academies, whose members were part of a transnational Republic of Letters (see Volume 4), after 1800 science and mathematics took place within the context of modern nation states.

A CENTURY OF INVENTIONS

With respect to the fruits of mathematical activity, the long nineteenth century was both highly productive but also, toward its end, deeply concerned with consolidating new knowledge. Just as in the history of physics, one often finds references to the mathematics of this period as classical as opposed to the modern style that began to emerge after 1900. These categories are necessarily vague in many respects, but still they prove useful when trying to understand the broad sweep of change. The massive body of knowledge that accumulated by 1914—what came to be called classical mathematics—consisted essentially of four fields. Two of these—number theory and geometry—already had deep roots in antiquity, whereas a third—algebra, as in the theory of algebraic equations—was first cultivated in non-Western cultures and only came to Europe during the late Middle Ages. The fourth field was analysis or infinitesimal calculus, invented toward the end of the seventeenth century and further developed during the century that followed. As highlighted in Volume 4, this newest branch of mathematics emerged along with differential equations, as both were used to advance another closely related field, namely classical mechanics (Gray 2021).

Major innovations took place within all of these areas of mathematics over the course of the century, yet, as described by Jeremy Gray in Chapter 3, there was a general trend toward purifying older results, which ran alongside the pursuit of newer ones. This new ethos of purism was a point of pride for the era's professional mathematicians, who advocated higher standards of proof, while they began to probe the foundations of all four fields of classical mathematical knowledge. This was thus the century that invented the notion of mathematical rigor in its modern sense.

Up until roughly 1800, most of the concepts encountered in mathematical texts represented idealizations of objects found in the natural world. In classical Antiquity, Greek mathematicians already conceived of idealized numbers and geometric figures. In the former case, this involved stripping away all properties

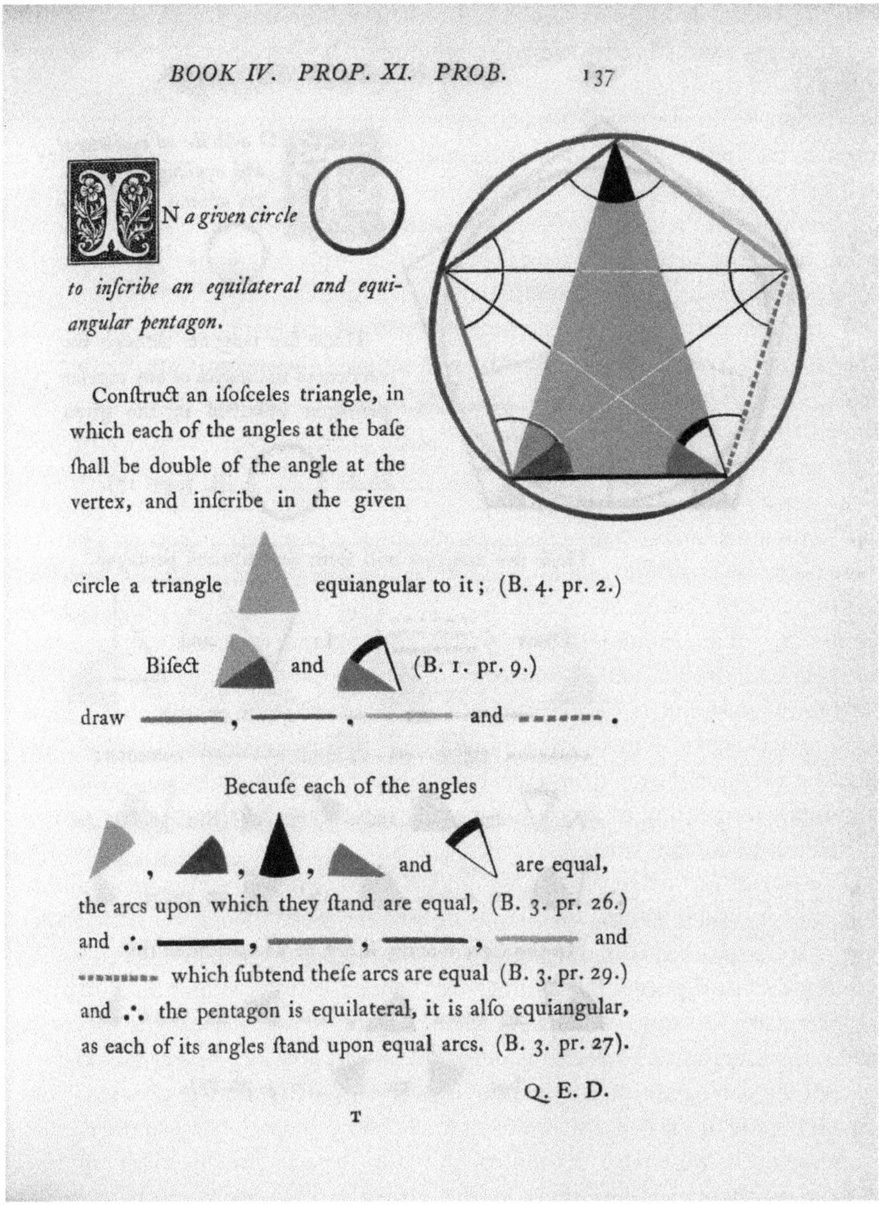

BOOK IV. PROP. XI. PROB. 137

IN a given circle to inſcribe an equilateral and equiangular pentagon.

Conſtruct an iſoſceles triangle, in which each of the angles at the baſe ſhall be double of the angle at the vertex, and inſcribe in the given circle a triangle equiangular to it; (B. 4. pr. 2.)

Biſect and (B. 1. pr. 9.)

draw , , and .

Becauſe each of the angles , , , and are equal, the arcs upon which they ſtand are equal, (B. 3. pr. 26.) and ∴ , , , and which ſubtend theſe arcs are equal (B. 3. pr. 29.) and ∴ the pentagon is equilateral, it is alſo equiangular, as each of its angles ſtand upon equal arcs. (B. 3. pr. 27).

Q. E. D.

T

FIGURE 0.3. A page from Oliver Byrne's color-coded edition of Euclid's *Elements*. Photo by SSPL/ Getty Images.

from a collection of things except for their multiplicity. In the latter, one imagines, for example, the properties inherent in a perfectly square object, one with exactly equal sides. This manner of thinking about abstract numbers and figures lent legitimacy to these concepts, since they mirrored familiar objects found in the natural world. Philosophers might still debate the actual ontological status of mathematical ideas, but few felt compelled to deny their existence altogether. Over the course of the nineteenth century, however, mathematicians began to extend the right of existence to other, seemingly impossible objects, a process affected by several unanticipated developments along the way.

Throughout history, mathematical knowledge has changed in subtle ways. This is not to say, however, that we should understand the history of mathematical ideas as a purely cumulative process. Some historians have even argued that mathematics has undergone the same kinds of massive paradigm shifts that Thomas Kuhn and others have pinpointed in the natural sciences and likened with political revolutions. Two striking instances of such momentous shifts in mathematics occurred, in fact, during the nineteenth century. The first came about when, after roughly two thousand years of efforts, mathematicians finally gave up the quest of trying to prove the validity of Euclid's famous parallel postulate. In modern texts on Euclidean geometry, one no longer finds the original formulation of Euclid's *Elements* but rather the version John Playfair published in 1795. Playfair's axiom asserts that for any line ℓ in a plane and point P not on ℓ there is a unique line ℓ' passing through P that does not intersect ℓ. For centuries, geometers had come up with what seemed like airtight arguments purporting to prove this parallel postulate. Yet, each such "proof" turned out to contain some unsuspected flaw.

This went on until the 1820s and 1830s, when the Russian Nikolai Lobachevsky and the Hungarian Janos Bolyai convinced themselves that all such efforts were doomed to failure for a very simple reason: the parallel postulate was not necessarily true. Put another way, Euclid had been right all along that his postulate—or some other equivalent version—was necessary in order to derive the theorems of classical geometry. Lobachevsky and Bolyai staked their case on the daring notion that we can conceive of an alternative geometry, one in which some of the basic theorems of Euclidean geometry, but also the parallel postulate itself, were false. It took roughly four decades for this shocking news to make mathematical headlines, but beginning in 1868 many mathematicians and even some philosophers had begun to see the light. Once this door fell open, it became more and more clear that the classical geometry of Euclid was merely a special case, one among an infinite multitude of other possible geometries.

The year 1868 saw the publication of two independent works that stirred great interest in physical questions connected with the foundations of geometry. Two years earlier, Bernhard Riemann passed away, leaving behind the manuscript from the habilitation lecture he delivered in 1854. Riemann had been none too eager to speak on this topic, but did so at the request of Carl Friedrich Gauss, who had long harbored doubts regarding the universal validity of the parallel postulate. Yet, he published nothing relating to these matters, fearing that doing so would drag him into disputes with philosophers, in particular those who adopted the views of Immanuel Kant. Within the Kantian system, the theorems of Euclidean geometry enjoyed a special epistemological status that transcended the laws of logic as well as the principles of empirical inquiry (Figure 0.3). Thus, for Kant, the parallel postulate constituted a synthetic a priori proposition, thus a higher form of truth. If Gauss anticipated that Riemann would address these issues, he was surely surprised to see that instead his former student approached the foundations of geometry from an entirely new perspective, namely by introducing what we today call n-dimensional Riemannian manifolds. This amounted to a profound generalization of the Gaussian theory of surfaces, though Riemann never published afterward on this topic.

Less visible than non-Euclidean geometry, but for modern mathematics even more profoundly influential was Georg Cantor's radically new theory of the infinite in mathematics. Cantor's philosophy of the infinite reflected his personal religious views, which need not concern us here (see, however, Dauben 1979). In fact, his starting point was a difficult problem in real analysis, not a question that arose from philosophy. In 1822, Joseph Fourier introduced the series that bear his name, claiming (incorrectly) that an arbitrary function was uniquely representable by an infinite sum of trigonometric functions, its Fourier series (see Chapter 3). Several German analysts proved various partial results, whereas Cantor began investigating the question of unique representation for functions with exceptional points. He was able to prove uniqueness even for functions with infinitely many exceptional points of a certain type. These infinite sets gave rise to a sequence of infinite derived sets that served as the prototype for what Cantor later called "transfinite ordinal numbers."

For Cantor, this marked the point of entry into his subsequent investigations of infinite subsets of the real number continuum, written $\mathbb{R}$ in modern notation and conceptually imagined as consisting of individual points on a continuous number line. This arithmetical continuum had long served as the fundamental meeting point for all four fields of classical mathematics, each of which was in some respect anchored to it. Cantorian ideas about infinite sets had little direct impact on geometry until after 1900, but eventually they dramatically transfigured a branch of geometry that had long lingered in the background. This was topology, which originally arose from considering properties of curves and surfaces under continuous deformations.

Part of the difficulty in studying such transformations involved the notion of continuity itself, long taken for granted until various vexing phenomena reared their heads toward the end of the nineteenth century. One of these was "Weierstrass' monster," a continuous curve that was so jagged one could not find a single point on it that possessed a tangent line. Another was Giuseppe Peano's invention of a continuous mapping of a line onto a square. This seemed hardly believable until Hilbert showed how to construct such a mapping geometrically (see Figure 4.5), while noting that his construction yielded another example of a Weierstrassian monster. Cantor's most famous example of a pathological set was produced by successively removing the open middle thirds from closed intervals, starting with [0,1]. Repeating this process infinitely often produces the Cantor set C, a totally disconnected set of measure zero (since the sum of the removed middle thirds is unity). Most startling of all, the set C is uncountable: it has the same cardinality as the entire continuum of all real numbers!

To what extent Cantorian set theory played a revolutionary role in the history of mathematics remains debatable, but it clearly helped set the stage for "modern mathematics." Several other developments over the course of the nineteenth century also entered this story, and some deserve mention here. The equation $x^2 + 1 = 0$, for example, has no solutions, unless one decides to admit a so-called imaginary number, $i = \sqrt{-1}$, since $i^2 = -1$. Such imaginaries had long led a kind of shadow existence in algebra, but C.F. Gauss and his contemporaries came to recognize their importance for higher algebra and number theory. Their utility is immediately apparent from the fundamental theorem of algebra, which pertains to the theory of equations (see Chapter 3 in Volume 4). The simple example above shows that not every equation has solutions in the real numbers $\mathbb{R}$, but in fact *every* algebraic equation has *all* its solutions in the complex numbers $\mathbb{C}$, which is a simple extension of $\mathbb{R}$ obtained by adjoining the number i: $\mathbb{C} = \{a + bi \mid a, b \in \mathbb{R}\}$. More precisely, the fundamental theorem asserts that a polynomial of degree n with complex coefficients has exactly n solutions (counting multiple roots) in $\mathbb{C}$. In modern parlance, one says that $\mathbb{C}$ is an algebraically closed field. Notice that this result has the character of a pure existence theorem: it says nothing whatsoever about how to find the solutions, which is a different and very difficult problem altogether. This example represents part of the shifting character of mathematical inquiry as described in Chapter 3. It illustrates how mathematics began to acquire two hallmark features of modernity: abstraction, and its companion, generality.

The admission of complex numbers spawned entirely new fields of research, in particular complex analysis, i.e., the study of functions of a single complex variable (Bottazzini and Gray 2013). By mid-century, this new theory of functions attracted numerous researchers, including such stellar figures as Cauchy, Riemann, and Weierstrass. Since every complex number can be written in the form $a + bi$, it can be represented uniquely by the point (a, b) in the complex plane. Exploring this new terrain, mathematicians quickly came to realize the many advantages of working with theories in $\mathbb{C}$ rather than restricting to $\mathbb{R}$. On the other hand, the former possibility of visualizing curves and surfaces by means of analytic geometry in $\mathbb{R}^2$ or $\mathbb{R}^3$ was no longer a simple matter. A complex curve in $\mathbb{C}^2$ (the counterpart to a real curve in $\mathbb{R}^2$) corresponds to a Riemann surface in $\mathbb{C}^2$, which is a four-dimensional space. Felix Klein and others sought to describe the properties of Riemann surfaces within the confines of $\mathbb{R}^3$, which involved fictitious intersections and other puzzling phenomena (see Figure 0.4). In Berlin, Weierstrass promoted an alternative non-geometric approach to complex function theory, which dominated the field up until around 1900. In 1912, Hermann Weyl finally presented a rigorous version of Riemann's theory and showed that it was mathematically equivalent to the Weierstrassian approach.

Thus, complex analysis gradually emerged as a fundamental field within what came to be known as classical mathematics. By the turn of the century, however, David Hilbert began to advance a new and more abstract approach based on modern axiomatics. The roots of Hilbert's mature formalist views trace back to this early work, but especially his axiomatic studies on the foundations of geometry. In his *Foundations of Geometry* (1899), Hilbert set forth five groups of axioms, each providing a new layer of structures and theorems, without however making any direct claims about the ontological status of the objects under discussion. Points, lines, planes, etc. were, as always before, the fundamental objects, but Hilbert made very clear, these were merely names. The axioms described the properties they satisfied—that, for example, two distinct points determine a line—but the words point and line had no

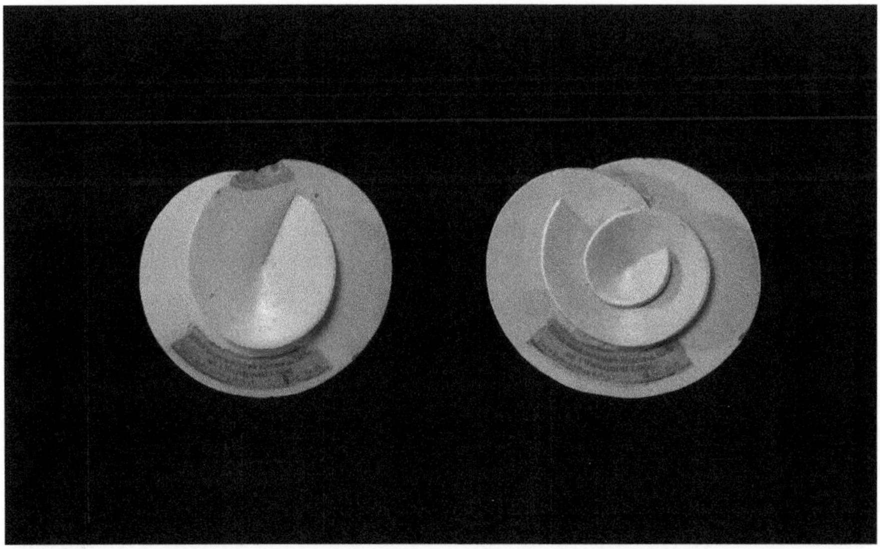

FIGURE 0.4. Two models of Riemann surfaces made by the company of Martin Schilling in Halle, Germany, *c.* 1900–10. © Collections de l'Institut Henri Poincaré.

further meaning or importance than this. As Hilbert once said, one could just as well speak of tables, chairs, and beer mugs so long as these objects fulfilled the requisite axioms.

In 1900, Hilbert delivered his famous lecture on mathematical problems at the Second International Congress of Mathematicians held in Paris, remembered especially for the 23 challenge problems he posed to his contemporaries. In his second Paris problem, Hilbert took his axiomatic program one step further by claiming it should be possible to give a direct proof that his axioms for the real numbers are consistent, meaning one could never deduce contradictory statements from them. No one, of course, had ever imagined this could occur, but Hilbert wanted more: he demanded a rock-solid proof of consistency. The first Paris problem involved a pair of conjectures—the continuum hypothesis and the well-ordering theorem—that were central for Cantor's theory of infinite sets. In raising these, Hilbert stepped forth as one of Cantor's strongest defenders. Set theory gained its first adherents among younger analysts, who found it an essential tool for their investigations, but after 1900 it would emerge as a major battleground for opposing philosophical disputes. The debates were enjoined by representations of three main competing camps: the logicists (Bertrand Russell and Gottlob Frege), intuitionists (Émile Borel, Henri Poincaré, and the Dutch topologist L. E. J. Brouwer), and formalists (Hilbert and Ernst Zermelo)

These developments were part of a broader trend that assimilated the classical theories of the nineteenth century by moving to a higher level of abstraction, one in which mathematical objects were introduced and defined by means of axiomatic systems. While it is difficult to pinpoint when the transition from classical to modern mathematics actually occurred, in broad terms the years from 1870 to 1914 mark a transition period during which many seeds of mathematical modernism were sown. A similar transition occurred in the physical sciences during this period, when new theories based on concepts like energy and electromagnetism and a plethora of new experimental findings gradually supplanted older, more simplistic theories. These shifts also led to major new developments in mathematical theory and praxis.

SCIENCE, MATHEMATICS, AND WORLDVIEWS

Throughout the nineteenth century, the respective claims of science and religion collided in the face of new theories regarding the age of the earth and the origins of life on it. Christian orthodoxy also faced new challenges posed by modern philologists, who analyzed the Bible critically as a historical document. As a young theology student, Charles Darwin read William Paley's influential *Natural Theology* (1802), which begins by invoking the analogy between the complexity of the natural world and that of an intricate watch made by a skillful watchmaker. At the time, Darwin found Paley's arguments compelling, but after spending five years as a naturalist exploring South America and studying its flora and fauna, he began to contemplate how species evolve through the principle of natural selection. His theory became famously controversial after 1859, when he published *On the Origin of Species*. In his posthumously published autobiography, Darwin summed up the intellectual journey he had traveled by writing, "The old argument of design in nature, as given by Paley, which formerly seemed to me so conclusive, fails, now that the law of natural selection has been discovered." Those who henceforth wished to reconcile the newest scientific theories with orthodox Christianity thus stood before a perplexing problem. No matter how they might attempt to solve this quandary, the implications of thermodynamics coupled with Darwinian evolutionary biology clearly required abandoning many features of the worldviews of the eighteenth century.

Another strand of influence arose a decade later, when Dedekind published Riemann's lecture "On the Hypotheses underlying the Foundations of Geometry." As Ivahn Smadja describes in Chapter 4, this rather mysterious text set off a string of debates among natural philosophers that resonated long afterward. Shortly before this time, Eugenio Beltrami had vindicated the earlier work of Lobachevsky and Bolyai on non-Euclidean geometry, and soon a number of leading mathematicians took up these promising ideas. In 1870, William Kingdon Clifford spoke before the Cambridge Philosophical Society about the physical implications of Riemann's theory. Clifford underscored the possibility that changing curvature effects in space might accompany or even account for matter in motion. Riemann's ideas were, to be sure, abstract and general, but at the end of his now famous lecture he also discussed the possibility that physical space might fail to have a Euclidean structure, thereby intimating that Immanuel Kant's position was untenable. Instead, he considered the global geometry of space to be an open question, as Albert Einstein later did as well. In 1917, in the context of relativistic cosmology, Einstein set forth a model of the universe in which space had the form of a giant three-dimensional sphere.

Shortly before the publication of Riemann's lecture, Hermann von Helmholtz independently developed an approach to the geometry of space similar to Riemann's, though their starting points were entirely different. Both nevertheless took an empirical stance, in opposition to the Kantian view that human space perception derives from the synthetic a priori character of Euclidean geometry. For Helmholtz's investigation, though, this empiricist stance was the essentially new aspect, as he emphasized by changing the word "hypotheses" in Riemann's title to "facts" in his own essay: "On the Facts underlying the Foundations of Geometry." What Helmholtz took to be the key axiomatic property of space was that it allowed for the free mobility of rigid objects. Indeed, one of the centerpieces of nineteenth-century physics was rigid body (continuum) mechanics. Only after 1900 did the opposing view—namely, that microscopic matter has a discrete rather than continuous structure—gain ascendency. A new twist on the theory of rigid bodies arose from relativity theory, which posits that a physical object will deform when accelerating. As a consequence, the classical notion of a rigid body no longer obtained.

Curvature was a magic word in the discussions about physical space that soon followed, though it already had more than one meaning in mathematical surface theory. Euler had proved the fundamental curvature theorems for surfaces embedded in Euclidean 3-space, whereas Gauss later showed how one can introduce a subtler concept, today called Gaussian curvature, based on measurements on a surface that are independent of any embedding. Riemann then found a way to generalize this approach to spaces (manifolds) in higher dimensions. In particular, he speculated that our universe might have the structure of a gigantic three-dimensional sphere, i.e., a space with a very small positive curvature. This terminology led many to imagine that such a Riemannian geometry necessarily presumed the existence of a four-dimensional space that contains this 3-sphere universe. Felix Klein once described the hopelessness of these types of discussions, citing a witty two-line ditty by Otto Blumenthal:

People barely grasp the case
Of curvature measure in space.
(Die Menschen fassen kaum es
Das Krümmungsmass des Raumes.)

Minkowski space offered a new framework for discussing paradoxical phenomena in space and time that fascinated the larger public. Popular discussions often suggested that relativity was a new form of science fiction connecting the fourth dimension with time travel. Long before this, in fact, speculation about various types of four-dimensional "hyperspaces" ran rampant. Particularly famous down to the present was the 1884 novella *Flatland: A Romance of Many Dimensions*, written by Edwin A. Abbott, an English schoolmaster. Although usually read as an amusing way to imagine a four-dimensional world, the author's main intention was to satirize Victorian society by viewing its citizens as geometrical objects in a two-dimensional world, Flatland. The story's narrator, "A Square," tells of a visit by a Sphere from a higher world, Spaceland. Living in a two-dimensional realm, though, he could only see Sphere's circular cross sections as this creature moved through Flatland. Abbott's witty tale helped spawn interest in visualizing four-dimensional objects, a problem that greatly fascinated the mathematician Charles Howard Hinton. In 1904, Hinton published *The Fourth Dimension*, in which he used color schemes to picture a tesseract, his name for the four-dimensional analog of a cube. Hinton's work influenced the daughter of logician George Boole, Alicia Boole Stott, who created impressive models of four-dimensional solids and coined the term polytope (Polo-Blanco 2008).

These types of interests and speculations naturally drew the attention of spiritualists and occasionally even leading scientists, like the Leipzig astronomer Karl Friedrich Zöllner. During the 1870s, Zöllner invited colleagues to his home, hoping to convince them by way of various séance experiments in the existence of a four-dimensional world. Ether physicists also sometimes speculated that electromagnetic phenomena resulted from vibrations in a higher four-dimensional space, an idea tentatively advanced by the American astronomer Simon Newcomb. By the turn of the century, public fascination with the fourth dimension made this a favorite theme among amateur philosophers, occultists, science-fiction writers, and artists (Henderson 1983/2014). These currents ran quite apart from the later widespread interest in Einstein's theory of relativity, which uncovered interconnections between

FIGURE 0.5. David Hilbert (1862–1943). Photo by Ullstein Bild / Getty Images.

space and time. Minkowski showed how these led to a new four-dimensional geometry, but here spacetime rather than space has four dimensions.

By the last decades of the nineteenth century, scientific positivism and secularism had begun to displace traditional philosophy and religion in many Western European societies. Nevertheless, many of the old metaphysical questions remained. A century earlier in his *Critique of Pure Reason*, Immanuel Kant circumscribed the realm of metaphysical knowledge to questions potentially open to human experience. Kantian conceptions still loomed large in Germany well into the twentieth century, just as they still did in 1872 when the distinguished physiologist Emil du Bois-Reymond delivered his famous lecture, "On the Limits of Science." He began by citing Kant's famous dictum—a favorite of another famous citizen of Königsberg, David Hilbert—that in any particular theory of nature one can find exactly as much science as the mathematics found in it. As for the limits of science, du Bois-Reymond spoke of the fiction raised by "Laplace's demon,"—the idea of total mathematical determinism from known initial conditions—claiming that humankind can never know the true nature of matter or the forces that act on it. His closing words, *ignoramus et ignorabimus* (we do not know and will not know), echoed for decades long afterward as a symbol of *fin-de-siècle* pessimism and became a kind of watchword in elite circles of Germany society.

Hilbert, who took issue with this general gloom, became after 1900 a leading voice for scientific optimism, although without addressing any of the questions du Bois-Reymond had declared unanswerable. Instead, in his Paris lecture, Hilbert wanted to take the Kantian motif a step further by claiming that mathematical knowledge has no boundaries. In setting forth this view, he reminded his audience how mathematicians had at last resolved two fundamental problems bequeathed upon them by the ancient Greeks, namely the status of the parallel postulate in Euclidean geometry and the impossibility of squaring the circle. These achievements illustrated what Hilbert saw as a general characteristic of mathematical problems, namely that every well-posed question in mathematics could be answered with finality. Indeed, he regarded this as a kind of moral credo for the mathematician: "the conviction in the solvability of every mathematical problem is a powerful incentive to the worker. We hear within us the steady call: there is the problem, seek its solution. You can find it by pure reason, for in mathematics there is no *ignorabimus*" (Gray 2000: 248).

Eight years later, speaking at the International Congress held in Rome in 1908, Henri Poincaré spoke about the future of mathematics in far less sanguine terms, warning in particular about the influence of Cantorism (Gray 2013: 118). In the meantime, a storm of controversy over Cantor's theory had broken loose, partly due to various logical paradoxes involving ill-defined sets. Even more controversial, however, was the argument Ernst Zermelo used to prove Cantor's claim that one could give every set a well ordering. To show this,

Zermelo invoked a seemingly obvious principle known as the Axiom of Choice (AC). This asserts that if given an arbitrary collection of sets, then we can freely pick out one element from each. This idea of "choosing" can be highly misleading, however, because if there are infinitely many sets, AC allows us to make these choices "all at once," that is independent of time. For finitely many sets, one can define an iterative process for choosing an element from each set in a given collection, but such a procedure may not be available for an infinite collection.

During the first decade of the century, Hilbert and Poincaré disagreed sharply over the legitimacy of AC in mathematical arguments. Three years prior to his unexpected death in 1912, Poincaré delivered a lecture on the status of set theory in Göttingen in the presence of both Hilbert and Zermelo. His skeptical views reflected a critique published by Jules Richard in 1905. This concerned so-called impredicative or self-referential definitions, which Poincaré took to be the core problem in set theory. Russell gave the following simple example to illustrate Richard's paradox. Let S be the set of natural numbers in N definable by no more than 100 words. Since the set S is finite, let m be the smallest number not in S: but then, by definition, m must be in S!

In Göttingen, Poincaré made very plain why he regarded Zermelo's entire line of argument nothing but a word game that lacked all mathematical substance:

> What should we think of the famous continuum problem? Can one well order the points of space? What do we mean by that? There are two possible cases: either one claims that the law of well-ordering can be formulated in finitely many words, then the assertion is unproved; Mr. Zermelo makes no claim to have proved such an assertion. Or else we admit the possibility that this law cannot be finitely expressed. But then I can no longer make sense of this statement; for me, these are only empty words. Herein lies the difficulty, and that is probably also the cause of the dispute over the almost ingenious theorem of Zermelo. (Rowe 2018: 200)

These strongly expressed reservations, coming in 1909 from the leading mathematician of the era, would hardly be the last word on Cantorian set theory. They reflect, however, the strong resistance it faced toward the end of the long nineteenth century.

HEAT, LIGHT, AND THE NEW FIELD PHYSICS

In 1800, the mathematical description of the natural world was resolutely based on Laplacian determinism. A knowledge of the positions and states of motion of the various kinds of substance—ponderable matter, light, heat, electric particles—combined with the laws that governed their interactions, would lead to a knowledge of all future states. The model here was celestial mechanics, especially the known planets in our solar system. In his five-volume *Traité de mécanique celeste,* published between 1798 and 1825, Pierre-Simon Laplace attempted to prove the long-term stability of the solar system. During these years, he also published major works on probability theory, including *A Philosophical Essay on Probabilities* (1814). This contains a famous passage, alluded to above as "Laplace's demon," which captures the "divine clockmaker" worldview. Laplace imagined an intellect capable at a certain moment in time of grasping the position of all particles in the universe and the forces acting on them, but also able to subject this data to mathematical analysis. For such a being, he asserted, the future as well as the past would lie open before its eyes. By mid-century, however, Laplacian physics had given way to other approaches. One of these was statistical mechanics, which is rooted in classical thermodynamics and the notion of energy conservation.

Even before Newton's time, classical mechanics had evolved alongside a search for conservation laws, statements pertaining to something that remains unchanged when changes in nature take place. Descartes proposed that "quantity of motion" (ms, mass times speed) was always conserved, a hypothesis that came under attack when Leibniz promoted *vis viva* (mv^2) instead. Newton's *Principia* underscored the distinction between velocity and speed by making clear that force-free motion led to uniform speed along rectilinear trajectories. If the direction of an object changes, then its velocity differs, though its speed may still be constant. Thus, velocity v refers to direction as well as quantity; in modern terminology, v is a vector, whereas s is a scalar quantity. This distinction was important in formulating the law of energy conservation in mechanics, which states that the total energy of a system—kinetic energy ($\frac{1}{2}mv^2$) plus potential energy—remains constant over time. In 1811, Joseph-Louis Lagrange gave an elegant proof of this in the second edition of his famous *Mécanique Analytique*.

Mechanical energy, though, proved to be only one of several forms. By the middle of the nineteenth century, scientists had amassed compelling evidence that whenever one form of energy is transformed to another the total quantity remains conserved. This finding, known today as the first law of thermodynamics, effectively launched a new field of fundamental research by introducing the concept of entropy in physical systems.

Lagrange's sophisticated mathematical methods set the modern standard for classical mechanics, which would provide a firm foundation for the discipline. By the last decades of the nineteenth century, however, two new physical theories arose as potential competitors for traditional mechanical ideas: thermodynamics and Maxwellian electrodynamics. As noted, thermodynamics emerged around the middle of the nineteenth century when the general concept of energy, but in particular heat energy, came into

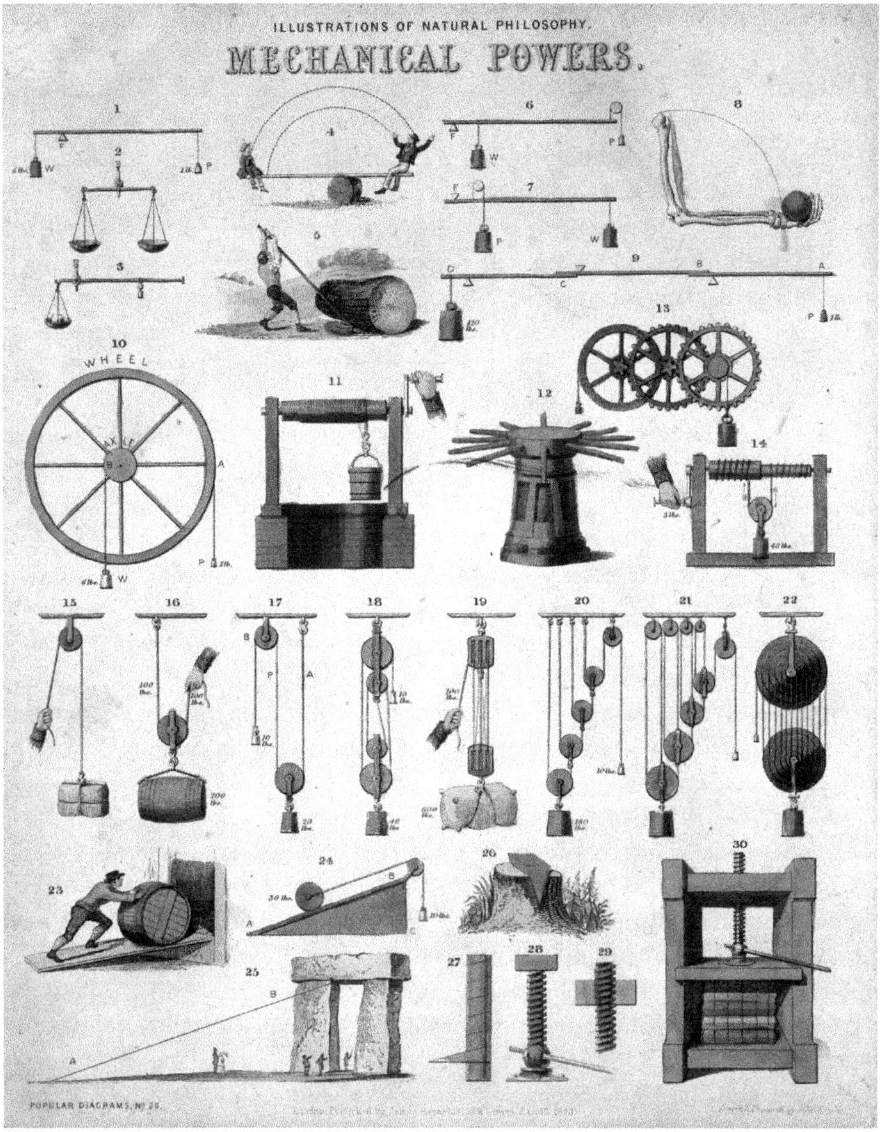

FIGURE 0.6. Pictures illustrating the Five Types of Powers in Classical Mechanics. Photo by SSPL / Getty Images.

focus, whereas electrodynamics began in the 1830s with Michael Faraday's studies of electrical and magnetic forces. Although both of these new fields differed significantly from classical mechanics, many leading physicists were long convinced that these new theories were at bottom compatible with mechanical principles. Maxwell, in particular, saw no inherent contradiction between his field theory of electromagnetism and Newton's theory of mechanics.

In Chapter 5, Scott A. Walter explains how thermodynamics emerged from an analysis of the efficiency of heat engines, which led to a deeper understanding of heat energy. In his *Reflections on the Motive Power of Fire*, Sadi Carnot analyzed the principles underlying an ideal heat engine that operated at maximum efficiency. Although little noticed during his lifetime, this book turned out to be a pioneering contribution to thermodynamics that marked a first step toward the concept of entropy, later introduced by Rudolf Clausius and William Rankine.

In 1850, Clausius formulated a form of the first law of thermodynamics asserting that the amount of work performed by a mechanical heat engine is equivalent to the heat consumed and conversely. Clausius's law rules out the possibility of constructing a perpetual motion machine of the first kind, which would produce work without requiring an energy source. This principle stands in close relation to the law of conservation of energy, which states that the total energy in an isolated physical system remains constant over time. Clausius also formulated the second law of thermodynamics—the assertion that the entropy of a physical system increases with the passage of time (for details, see Chapter 5). This law implies that physical events follow the "arrow of time" and are essentially irreversible. The closer a system's entropy approaches its maximum, the greater the disorganization of its material structures. Thus, while the total energy of an isolated system remains constant, it will tend to dissipate into heat energy over time. By the latter part of the century, the implication that our universe is spiraling toward "heat death" became part of the cultural discourse debated by *fin-de-siècle* pessimists.

Statistical mechanics arose from the closely related kinetic theory of gases. Already in his *Hydrodynamica* (1738), Daniel Bernoulli presented the basic notion underlying this theory, according to which heat and pressure arise from the collision of gas molecules. Clausius was the first to provide a theoretical argument for the kinetic theory based on a statistical law, which Maxwell elaborated further by invoking a model of gas particles that collide with one another randomly. Not long afterward, in the 1870s, Ludwig Boltzmann developed statistical mechanics as the underlying model to account for thermodynamic phenomena. Laplacian determinism, once the dominant paradigm for understanding the physical world, had long before fallen out of favor, weakened by the advent of Fourier's theory of heat conduction and Fresnel's wave theory of light. Classical mechanics needed to adapt even further once the role of randomness entered the picture in a decisive way.

Maxwell and Boltzmann saw no fundamental conflict between thermodynamics and mechanics; both appealed to probabilistic arguments to explain thermodynamic principles. Ludwig Boltzmann tried to reconcile entropy with classical mechanics. Entropy's "arrow of time" seemed, in fact, to contradict the fundamental laws of mechanics, which are time-symmetric; thus, one can replace the time t by $-t$ without affecting the equations of motion. These issues were debated throughout the period before a consensus was finally reached. In the meantime, others, most notably Wilhelm Ostwald, hoped to use energy conservation as a new basis for explaining classical mechanics. Like earlier representatives of German *Naturphilosophie*, Ostwald and other so-called energeticists rejected "scientific materialism" founded on the hypothesis that matter is composed of elementary corpuscles, atoms or molecules. This hypothesis served as the basis for the kinetic theory of gases, which linked mechanics with thermodynamics. For most energeticists, energy was applicable to mental phenomena as well as the material world. In a broader sense, energetics reinforced the influential views of Ernst Mach, who argued from a positivist perspective that since no empirical evidence supported the existence of atoms physicists should reject theories based on atomism.

Another short-lived theory, first proposed by William Thomson, emerged during the 1870s and 1880s. This was his vortex atom theory, an attempt to account for material structures by way of vortices in a continuous ether. This was but one of many ether theories, which were much in vogue among physicists circa 1900. Thomson's theory intrigued his colleague Peter Guthrie Tait, who began a systematic classification of knots to lend it support. In the course of doing so, Tait set forth three conjectures in knot theory that turned out to be correct, though it took another century to prove them. A weak point in the vortex atom theory stemmed from the fact that it could not account for ponderable mass.

An alternative theory that seemed more promising arose from electrodynamics, in which electric particles had an electromagnetic mass. This raised the question whether the entire mass of these particles was electromagnetic, in which case ordinary matter might consist of charged particles in a surrounding ether. Electromagnetism was another area of physics with ties to the notion of an ether. Advocates of wave theories of light had long appealed to a luminiferous ether as the medium by which light propagated through space. Maxwell's theory predicted that light waves were merely the visible part of the electromagnetic spectrum, a startling result that Heinrich Hertz confirmed in the late 1880s by conducting a series of experiments that led to the detection of radio waves. By the 1890s, Maxwell's theory and the existence of electromagnetic waves stood as firmly established, though the physical foundations for both remained unclear. Various proposals and claims for a new ether physics circulated, whereas Hertz himself voiced the opinion that "Maxwell's theory is Maxwell's equations." Indeed, the latter would survive far longer than the ether, though the mathematical formalisms used to express these equations were still in the process of being invented.

The Dutch physicist Hendrik A. Lorentz advanced an influential electron theory in the 1890s, and in 1897, Joseph J. Thomson discovered negatively charged electrons in a celebrated series of experiments with cathode rays. A consequence of Lorentz's theory is that an electron's electromagnetic mass increases along with its velocity, in other words, there was a relationship between mass and energy. This idea thus predated Einstein's theory of relativity, though Lorentz's approach to electrodynamics stood outside the then fashionable electromagnetic research program. In 1905, Einstein dismissed the ether as an unnecessary physical construct, while deducing his famous formula $E = mc^2$ from fundamental principles. This was a startling result, since it implied that even an extremely small quantity of matter m is equivalent to a huge amount of energy E, though no one, including Einstein, could then imagine there might be a way to tap into it.

Einstein's theory of special relativity, along with the quantum theory launched by Max Planck in 1900, would later emerge as fundamental for modern physics (see Chapter 5 of Volume 6). The earlier theories that once made claims to supplant the mechanical worldview of the eighteenth century—thermodynamics, electromagnetism, and ether physics—were either assimilated into this new framework or, in the case of ether theory, simply became obsolete. By the first decade of the twentieth century, relativity and quantum physics had already won over several leading adherents.

A key turning point for quantum theory came in 1913, when Niels Bohr introduced a new model for the atom based on Planck's constant. Bohr's provisional solution to probing the inner structure of the atom provided a fruitful model for subsequent research, though its inadequacies began to become apparent in the early 1920s. In the meantime, Einstein's generalized theory of relativity passed two key experimental tests proving its superiority over Newtonian gravitational theory.

For mathematicians, the success of general relativity opened the way to numerous new investigations, picking up where Riemann left off, as it were. This new spacetime physics raised the tempting possibility of combining gravity and electromagnetism in a unified field theory. These breakthrough developments thus laid the groundwork for much new physical research that would require new or improved mathematical methods of relatively recent vintage. Indeed, these shifts in the world of physics soon had major repercussions for mathematics. Following the death of Minkowski in 1909, Hilbert's attention turned more and more to physical problems for inspiration, hoping to employ axiomatic methods in a quest to clarify fast-breaking developments, especially in relativity and quantum theory (for which, see Chapter 5 of Volume 6).

MATHEMATICS INTERSECTING SCIENCE AND TECHNOLOGY

The period from 1870-1914 has commonly been regarded as marking the Second Industrial Revolution, symbolized by new forms of energy, production, and transportation. Whereas the First Industrial Revolution led to the growth of industries based on coal, iron, and railroads, this second period witnessed the development of new technologies built around electricity, steel, and petroleum-powered vehicles. In Chapter 6, June Barrow-Green and Tony Royle survey these and other technological innovations of the long nineteenth century, up to and including research and development in aeronautics during the First World War. Just as Watt's coal-burning steam engines radically altered the textile industry, so did the invention of the Bessemer process in steel production transform urban architecture by making it possible to build the modern skyscraper. Moreover, the availability of cheap steel led to large-scale bridges and ships and a vast expansion of the railways.

Probably the most apparent and dramatic change came about through the electrification of cities, which transformed modern social life dramatically. Before the advent of public electricity, nearly all types of human activity took place by sunlight; people still used candles and gas lamps to light their homes. Today such a life seems unthinkable, and while some parts of the world remain without access to a power grid, many feel deprived if their internet browser reacts too slowly. This dramatic technological revolution has far older scientific roots, but mathematics became more and more entangled with electrification. As the understanding of electromagnetism deepened, new mathematical conceptions were invented, some of which proved well adapted to new physical theories through a complex interplay of ideas. By the end of the long nineteenth century, this dynamic process led to a stable body of knowledge and a new group of scientifically trained technologists, the electrical engineers.

Perhaps the earliest quantitative finding of interest for electro-technology was Ohm's law, first presented by Georg Simon Ohm in 1827. This expresses the current I in a conductor in terms of the voltage V and resistance R by the simple formula $I = V/R$. Ohm's original formulation was inconsistent with established principles of electrostatics, but in 1849 Gustav Robert Kirchhoff showed how to reconcile it with the latter. The origins of electrostatics go back to Coulomb's law (1785), which describes the force acting between two charged bodies separated by a distance r. Like Newton's universal law of gravitation, this is a so-called inverse-square law, but with the difference that the force F is attractive for unlike charges and repellent for like charges. Newton's law takes the form $F = Gm_1m_2/r^2$, where F is the mutual gravitational force acting on two bodies with masses m_1 m_2. Coulomb's law describes the mutual electrical force between two charged bodies: $F = kq_1q_2/r^2$, where q_1 q_2 are the signed magnitudes of the charges. The larger significance of this law stems from the fact that from it one can prove Gauss' law, which relates the distribution of electric charge to the resulting electric field. In 1813, Carl Friedrich Gauss published his flux theorem, though not in an electrical context; it is taught today as one of the four Maxwell equations in the theory of electromagnetism.

Coulomb's law and other similar inverse-square laws played a major part in efforts to unlock nature's secrets. This trend continued even in electrodynamics through the work of Wilhelm Weber, who around mid-century derived a velocity-dependent generalization of Coulomb's law. A longstanding objection to these Newtonian-inspired force laws, however, was that such theories implied "action at a distance." This meant that, when considering two bodies separated in space that somehow interacted with one another, one ignored not only the medium transporting the action but also the time of transmission. In effect, proponents of "action at a distance" followed Newton in accepting instantaneous transmission, although already in the seventeenth century the Danish astronomer Ole Rømer had shown that light traveled at a finite speed. Laplace claimed that if gravity traveled at a finite speed, then this had to be several orders of magnitude faster than light.

With regard to electrical phenomena, a very different type of theory arose, however, inspired by the work of a natural philosopher who knew nothing about higher mathematics, namely Michael Faraday. Much of Faraday's research involved experiments aimed at understanding the interplay between electricity and magnetism based on what he called "lines of force." In the hands of his Scottish successors—William Thomson (Lord Kelvin), Peter Guthrie Tait, and James Clerk Maxwell—a radically different theory of electromagnetism eventually emerged based on the notion of a physical field. One of Faraday's inspirations concerned a rotary force in the magnetic field, a notion that appealed to Maxwell but not to Thomson. Maxwell had the deepest respect for Faraday's work, though Thomson's was surely just as influential. The latter developed a theory of electric conduction analogous with the flow of heat in an infinite solid. Maxwell saw this as a true breakthrough for field physics, since what Thomson showed was that the equations relating to the electrical field, a theory based on action at a distance in the form of Coulomb's law and its extension by S. D. Poisson, were identical with those derived from heat theory based on interactions between contiguous bodies. He later advanced a different analogy between electrical phenomena and the displacements transmitted through elastic bodies. Thomson was one of the last major natural philosophers who adhered to a mechanistic worldview. He believed that all forces in nature resulted from matter in motion. Maxwell also constructed a mechanical model that could account for electrical and magnetic phenomena (see Chapter 5), but he eventually abandoned this approach.

In 1873, Maxwell presented the essential ideas underlying his new conception in the monumental 2-volume study, *A Treatise on Electricity and Magnetism.* In this work, he concluded that light was actually an electromagnetic wave, or put in more modern terms, it represents that part of the electromagnetic spectrum visible to the human eye. He was able to draw this conclusion from the equations that today bear his name. Maxwell's equations, perhaps more than any others in

FIGURE 0.7. The apparatus used by Heinrich Hertz to generate and detect electromagnetic waves. Interfoto / Alamy Stock Photo.

physics, have had a deep and far-reaching impact on society. Without them and the resultant understanding of electromagnetic waves, we would be living in a world without radio, radar, television, or other wireless technologies. Unfortunately, Maxwell died in 1879, seven years before Heinrich Hertz was able to generate and detect electromagnetic waves in his laboratory in Karlsruhe (Figure 0.7). Hertz, too, died young, and after reading Righi's obituary of him, the Italian engineer Guglielmo Marconi began experimenting with wireless telegraphy.

Part of this complex story concerns these four famous equations due to Maxwell, which occupy a central place in all modern textbooks on electromagnetism. Although nowhere near as iconic as Einstein's famous formula $E = mc^2$, this is partly because the Maxwell equations are expressed in an unfamiliar mathematical language that uses concepts from vector analysis, such as curl and divergence. In fact, the modern version of this language only emerged well after Maxwell's lifetime, so this presents an important case of a mathematical formalism that developed in part by adapting to a new physical theory. In his *Treatise*, Maxwell wrote twenty equations instead of just four.

Hertz and, independently, Oliver Heaviside later gave them their modern form: the Hertz-Heaviside equations. By so doing, Heaviside helped to canonize today's vector analysis, a mathematical formalism he and J. W. Gibbs independently co-invented by distilling it from older theories, especially the four-dimensional formalism of quaternions set forth by William Rowan Hamilton.

Maxwell's *Treatise* played a major role in shaping subsequent developments precisely because its author freely used quaternions throughout much of the text. Indeed, Gibbs and Heaviside both first learned about quaternions from reading Maxwell and only afterward turned to Tait's *Elementary Treatise on Quaternions* (1867), which also stressed applications to physics. Since electromagnetism was concerned with phenomena in three-dimensional space, both must have felt inclined to use a mathematical formalism that utilized three components rather than four. Heaviside, as a practicing electrical engineer, reflected the pragmatic attitude of a highly creative, self-taught mathematician. Another fascinating figure from this period was Charles Proteus Steinmetz, a German immigrant to the United States who helped familiarize American electrical engineers with the mathematics behind systems operating on alternating current. Gibbs, on the other hand, came from the academic world and fully considered other systems, in particular that of H. G. Grassmann, which eventually gave methods for unifying the treatment of vectors with other more general mathematical systems. Still, Maxwell's *Treatise* was the true source of inspiration for the two men credited with inventing vector analysis; both came to this branch of mathematics through their interest in the physical sciences.

VISUALIZING AND REPRESENTING MATHEMATICS

When J. C. Maxwell first began studying Faraday's lines of force, he wanted to grasp the physical phenomena qualitatively. He found a way to do so by means of an analogy between electrical currents and the properties of an incompressible fluid. Ten years earlier, William Thomson had reasoned similarly by establishing an analogy between Fourier's theory of heat propagation and Poisson's treatment of electrical charges in equilibrium based on Coulomb's law in electrostatics. These Scottish physicists were both very adept at creating geometrical models that enabled them to visualize or imagine essential properties of electricity and magnetism. In trying to understand their interrelations, Maxwell had to abandon his first model and build another mechanical model with a honeycomb structure (see Chapter 5). Through a long complex process of trial and error, accompanied by visual thinking, he ultimately arrived at the mathematical equations found in his *Treatise on Electricity and Magnetism*. By the time his intellectual journey ended, however, the models had all disappeared, and yet they had accompanied him a good bit of the way.

In Chapter 7, Anja Sattelmacher and David E. Rowe focus on the role of physical geometrical models and their builders, stressing their importance as pedagogical aids. Like many other mathematical developments discussed in this volume, the roots of this story go back to Paris before moving on to various institutions of higher education in Germany and England, and then finally spreading beyond Europe, in part through international expositions and mathematical conferences. A great variety of geometrical models, many of them made in plaster, eventually went on public display, though usually as part of much larger collections that included all kinds of mathematical instruments. A special focus fell on the challenges of representing and visualizing exotic geometrical objects, such as the Riemann surfaces mentioned earlier (Figure 0.4).

Architects and construction engineers had long made use of physical models in their work. In fact, already in the late Renaissance, one finds examples of relief perspectives built for stage productions; these were three-dimensional models constructed by employing perspective principles. Such techniques likely played a role in courses taught at the École polytechnique, though its curriculum placed far more emphasis on descriptive geometry, which involved systematic use of parallel projections to convey the properties of three-dimensional figures on paper. This led to a major advance in teaching methods for technical drawing. One of Monge's students, Jean-Victor Poncelet, generalized these methods and opened the way for a new discipline: projective geometry. In effect, Poncelet's approach turned descriptive geometry into a special case by replacing parallel projections with central projections, a technique Italian Renaissance painters had used to convey a sense of visual perspective. In projective geometry, one introduces "ideal points" at infinity where parallel lines would intersect. Most of the models designed by French mathematicians, however, dealt with the properties of well-known objects from classical geometry, such as spheres, cylinders, or cones. In projective geometry, on the other hand, one cannot distinguish between a sphere and an ellipsoid, whereas a cylinder is just a cone whose vertex happens to lie at infinity.

An important collection of geometric models was created over decades at the Conservatoire National des Arts et Métiers (CNAM) in Paris, where Théodor Olivier taught beginning in 1838 (Brechenmacher 2022). He was also a founder of the École centrale des arts et manufactures, which opened in 1829; there he held the professorship for descriptive geometry and mechanics. Olivier became famous for building numerous models of ruled surfaces, which he made by attaching silk threads to a brass frame. One can imagine a ruled surface by thinking of an object swept out by a line that moves through space or, alternatively, by contemplating a surface composed entirely of straight lines, for example a cylinder. Olivier's models were especially popular in the United States, where Harvard, West Point, and Union College placed them on display. A similar collection was purchased in 1872 by the Science Museum in London, which ordered them from the Paris manufacturer Fabre de Lagrange. Four years later, they went on display when the museum hosted the first major international exhibition of scientific instruments, with around 4,500 objects.

Even though the Olivier models became particularly famous, they formed only a small part of the collection at the CNAM, most of which were used to impart technical knowledge. Libre-Irmand Bardin, also a student of Monge, contributed a large number of models and instruments related to stone cutting, gearing, and testing building materials. Bardin's student and collaborator, Charles Muret, built on these artifacts and marketed them through the publishing firm of Charles Delagrave as the Collections stéréotomiques de Muret. Thread models were particularly suitable for displaying ruled surfaces, but since they were fragile and by no means easy to construct, other materials were often preferred, such as cardboard, wood, or plaster. Julius Plücker designed a series of models in the 1860s that were manufactured using a heavy metal, either zinc or lead. Chapter 7 underscores how the choice of materials for building a model often had implications for its use as a teaching tool.

Although Plücker's models of so-called complex surfaces proliferated throughout Germany and even beyond, they have only recently been brought back to life, so to speak, as part of renewed interest in scientific objects (Rowe 2022). Arthur Cayley and other leading British mathematicians became interested in these models, which made it possible to visualize various types of so-called complex surfaces, exotic geometric objects that arise in the new field of line geometry. Plücker designed models for some 30 cases, which represent only a small fraction of all the various possibilities. After he spoke about them at a conference in Nottingham, he sent a set of boxwood models to the London Mathematical Society (LMS), which the society then put on display at the South Kensington Exhibition in 1876.

These models (see Figure 0.8) represent surfaces enveloped by different collections of lines in space, all of which intersect a fixed line (usually at infinity, the axis of a family of parallel planes). Since, as finite beings, our eyes cannot see something infinitely far away, projective geometry appeals to the mind's eye. Geometers of the nineteenth century became very adept at picturing projective objects by way of mental extension. Such a facility is required to appreciate many of the more sophisticated geometric models from the latter half of the century, namely those representing projective objects by revealing only a portion of the figure. Visualizing these *in toto* thus required that the viewer imagine how the object "passes through infinity."

By the turn of the century, many universities in Europe and North America had purchased models sold by companies in France as well as Germany. A particularly impressive collection came from the Sorbonne, where Gaston Darboux employed these geometrical artifacts as visual aids in some of his courses. His assistant, Joseph Caron, designed and built some eighty models

FIGURE 0.8. Plücker Models donated to the London Mathematical Society. © Science Museum Group.

made out of wood; many of these unique specimens are on view today at the Institut Henri Poincaré (IHP) in Paris, which opened in 1928. Nearly all the other models on display at the IHP, on the other hand, were commercial products found today in other collections as well.

Representation in the broader sense of public display also played a major part in the use of mathematical artifacts (Mehrtens 2004). Indeed, some of the mathematicians who promoted model making also took the lead in organizing exhibitions that centered on geometrical models and other symbols of their profession. These efforts gained momentum in Germany shortly after the founding of the German Mathematical Society in 1890. Three years later, a major mathematical exhibition took place in Munich, organized by Walther von Dyck. He organized this on an international scale with an extensive catalog that provided detailed information about the objects on display. Over one hundred exhibitors from ten countries took part in this event. Dyck later played a major role in the founding of Munich's Deutsches Museum, the largest museum of science and technology in the world.

Walther von Dyck's experience with model making dated back to his student days at the Munich Institute of Technology (Technische Hochschule or TH), which became a true hub for the production and distribution of geometric models in Germany. This began in 1875, when Alexander Brill and Felix Klein installed a workshop for students with facilities for designing and building prototypes of mathematical models. This activity proved so popular and successful that Brill began marketing many of these models through his family's publishing business in Darmstadt, run by his brother Ludwig. By the turn of the century, one could find collections of Brill models in virtually every major mathematics department in Germany and far beyond its borders as well.

Decisive for the success of Brill's undertaking were the many elegant plaster models in his collection (see Chapter 7). The aesthetic qualities of these artifacts was hardly left to chance, however, for although the prototypes were produced at the TH Munich, the casts for the models came from the workshop of Joseph Kreittmayr, a master craftsman who made plaster casts of antique statuary for the Bavarian National Museum. The students at the TH used a different modeling clay for their prototypes because plaster hardens very quickly and was therefore not easy to modify. Brill brought these master models to Kreittmayr, who then used them to produce negative molds for plaster casts.

Until 1899, when the firm of L. Brill sold its rights to Martin Schilling in Halle, the Brill family dominated this niche in the lucrative market for mathematical instruments, machines, and visual aids. Most of the geometric models in Dyck's catalog for the 1893 Munich exhibition came from the Brill collections. In that same year, American visitors to the Chicago World's Fair were able to view a collection of these very same plaster models in showcases. This was part of a larger exhibit financed by the Prussian Ministry of Culture to display the achievements of the German universities. Felix Klein engaged Dyck to organize a smaller collection of models and instruments for this purpose. Probably not many of the millions who visited the fair that summer took much notice, but a small group of mathematicians attended lecture demonstrations by Klein, who explained some features of the more complex objects. The Brill firm actually had this particular foreign market in view for some time; it had been regularly advertising its models in the *American Journal of Mathematics* since 1887. That too proved a success, especially on the east coast of the United States, where nearly all the universities and colleges owned collections of its models.

By the turn of century, the vogue for building and exhibiting geometrical models had passed its peak. Italian geometers, who dominated the field by the 1890s, had far less interest in visualization. Many of them worked on the forefront of axiomatic geometry, though intuitive methods would remain an important part of the Italian research tradition for decades to come. In the meantime, older models, when they survived, assumed the status of museum artifacts. The recent revival of interest in them forms part of the larger topic of visualization in mathematics after 1990 taken up in Volume 6.

CHAPTER ONE

Everyday Numeracy

Numeracy in the Age of Mass Education

ADRIAN RICE

The nineteenth century was the period when the current conception of the utility of basic arithmetic began to emerge. Although everyday numeracy had long been the ability to understand and compute with numbers, it was during this era that the fields in which these numerical skills could be applied increased dramatically, driven by rapidly evolving developments in technology, economics, trade and industry. Yet, while the notion of what was meant by a basic competence in the use and manipulation of numbers would have been largely understood throughout the nineteenth century, no single term existed to describe such a skill. It would not be until the word "numeracy" was coined in a 1959 report for the British Ministry of Education that the English language was provided with a term analogous to the meaning of "literacy"—itself a concoction of late-nineteenth-century American English. This chapter will thus discuss a concept for which no name existed at the time. It is therefore perhaps unsurprising that some basic parameters will be necessary before we begin.

For the entirety of the nineteenth century, everyday numeracy would have mainly comprised the ability to perform numerical calculations correctly. But since no understanding or appreciation of the underlying reasoning or logic was required, such mechanistic skills were largely considered to be beneath the higher concerns of true mathematicians. As Leibniz had lamented in 1685: "It is unworthy of excellent men to lose hours like slaves in the labor of calculation which could be relegated to anyone else if machines were used." Such sentiments were common among the educated elites in the eighteenth and nineteenth centuries—in 1847, an article in the *Dublin Review* observed: "We suppose it will hardly be disputed, that to speak the opinion of mankind, we must say that of all disgusting drudgery, numerical calculation is the worst: a combination of all the worry of activity with all the tediousness of monotony and all the fear of failure."

But numeracy was not all mere number crunching. Although nineteenth-century education in arithmetic continued the eighteenth-century emphasis on memory, rote-learning and reckoning, the demands of a changing society ultimately required increasing sections of the populace to be able to use, interpret and apply the results of numerical computations in an ever expanding variety of settings, from household management to commerce, demography to social reform. To understand this nineteenth-century expansion in the applicability of numeracy, we will attempt to answer a variety of questions. To what extent was education in arithmetic available at this time, and to whom? What form did this education take, and what skills were imparted by contemporary textbooks? What systems of weights, measures and currency were in place, and what efforts were made to facilitate their use? What other resources were available to facilitate everyday numerical computations? And to what extent were people exposed to data and numerical information in everyday life?

Clearly, the answers to such questions are highly location-dependent and, even if all the necessary historical sources were available, it would still be impossible to provide satisfactory answers on a global scale. However, an illustrative case study provides an excellent alternative. And since Britain was the world's foremost economic, industrial and military power throughout this period, it makes sense to use Victorian Britain as a geographical focus for this chapter, as developments there were often highly influential elsewhere around the globe, particularly in contemporaneous and former colonies, such as the United States of America. During the nineteenth century, the close cultural connection between Britain and the United States, which had existed since colonial times, remained strong. Thus, since basic numerical mathematics was a fundamental component of this shared cultural relationship, this chapter will focus on aspects of everyday numeracy in Britain and the United States in the nineteenth century, comparing and contrasting the provision, application, and manifestations of numeracy in the daily life of both cultures.

EDUCATION

The best place to begin in any survey of this kind is with a discussion of the provision of education in the rudiments of arithmetic and numerical calculation. Both Britain and America witnessed a tremendous expansion in the availability of school-level education during the nineteenth century, having begun the period with no systems of compulsory education in place. In early nineteenth-century Britain, such formal education as did exist was provided either by individual schoolmasters or, more frequently, by the church, particularly in "Sunday schools" which had begun to spring up since the 1780s. The establishment of the National Society for Promoting the Education of the Poor in the Principles of the Established Church in England and Wales in 1811—and its nonconformist counterpart the British Society for Promoting the Lancasterian System for the Education of the Poor—inaugurated an assortment of new institutions for the working classes. While the aims of these so-called "National" and "British" schools were mainly for religious instruction, their curricula also included tuition in basic arithmetic.

Although legislation had enforced the creation and funding of schools in every parish in Scotland since the seventeenth century, in England and Wales the government was slower to become involved in the provision and supervision of public education. But schooling for the general populace began to be supported at the state level during this period. From 1833 government money was allotted to the construction of schools for poor children, with other initiatives also beginning in that decade, such as the opening of the first Teacher Training College in 1839. For older children, the Grammar Schools Act of 1840 expanded the curriculum from classical studies to include science, literature and mathematics. Yet, despite these developments, even the provision of a rudimentary grounding in elementary arithmetic could not be taken for granted universally. In the 1840s, over 30 percent of British children still received no formal education at all and of those who did, the usual duration of their schooling was between one and two years. In the 1850s, a survey of 5319 schools revealed that less than half of the children in them were being taught arithmetic.

For those pupils who did receive an arithmetical education, the principal method for much of the century relied mainly on oral instruction. Although the children would often write on small slates, larger blackboards for the teachers were only introduced into classrooms in the first third of the century, becoming more common after around 1840. The teacher would recite (or copy out) definitions, rules, and examples, which the students would then write down verbatim. The main emphasis was on rote-learning, with recitation of multiplication tables being a prime example of such memorization. As the century progressed certain more enlightened schools and teachers introduced various improvements, such as using tangible objects to introduce ideas of quantity and operation and allowing children to explore simple examples before introducing more general rules and procedures. But it was more often the case that the subject was taught in a drill-like manner by teachers who had disliked the subject when pupils themselves, thus imparting no interest, understanding or enjoyment to a subject deliberately shorn of all meaning and relevance to the student.

FIGURE 1.1. This picture from *Baby's Own Alphabet* (1875) by Walter Crane, along with its accompanying verse, is illustrative of the experiences of many Victorian schoolchildren with arithmetical education. Photo by Jennifer Kennard/Corbis / Getty Images.

It was the Education Act of 1870 that marked a sea-change in the history of education in Britain, mandating the right to an elementary education for every child for the first time as well as making the government responsible for its provision. Further legislation in 1880 made education compulsory for children between the ages of 5 and 10, with certificates issued to certify the attainment of acceptable levels of basic skills including elementary arithmetic. With all elementary education becoming free from 1891, the school leaving age was raised to 11 in 1893 and again to 12 in 1899—although in Scotland, the leaving age had been 13 since 1872. At the opening of the twentieth century, the 1902 Education Act resulted in a rapid growth in the number of secondary schools, with over a thousand opening by 1914, including 349 for girls.

This expansion of the availability of education in Britain was reflective of larger international trends during the nineteenth century. In the previous century, Prussia had been one of the first countries to provide free and (largely) compulsory elementary education. In the first half of the nineteenth century, as part of Wilhelm von Humboldt's educational reforms, the mathematics taught in Prussian elementary schools ("Volksschulen") came to be viewed, not as mere arithmetic, but as an integral component of "Bildung" the comprehensive intellectual and emotional development of the individual. This attitude was often accompanied by progressive teaching methods, which sought to inculcate the practices of "calculating by thinking" and "self-directed activity." In the wake of the revolutionary events of 1848, however, the second half of the century saw the rise of less progressive ideologies, with the amount of arithmetic taught at elementary schools reduced and, after the unification of Germany in 1871, greater centralization of the school system. But this period also saw the growth of provision and state support of education for girls.

As in many of the German states, the development of education in nineteenth-century France was marked by a strong tension between the church and state, with the latter increasing in dominance throughout the period. In 1789, more than three-quarters of French women and half of French men were illiterate. And although Napoleon established a secondary school ("lycée") system in 1802, this did nothing to improve elementary education, which was largely provided by the Catholic Church. The "Guizot Law" of 1833 marked the beginning of a national state school system in France, with every local community required to provide an elementary school for its children. But the education was not free, it was confined mainly to boys and, in arithmetic, comprised mainly dictation and rote learning. The steady growth of industry and an affluent middle class throughout the century, together with a fall in the need for child labor, led to further reforms. By 1882, fees for elementary schools had been abolished, secondary education for girls had government support, and school was compulsory from the age of 6 to 13. By 1910, with the overall illiteracy rate at 4.2 percent, the French government exercised full control of the national curriculum and school examinations, as well as the training and certification of teachers.

Across the Atlantic, in the United States the story proceeded along similar lines. In the sixteen states of the Union in 1800, most schools paid little attention to the need for instruction in arithmetic. In common with their European counterparts at the same time, the majority of elementary schooling was provided either by a single schoolmaster or by religious bodies, and the ultimate aim was almost always for a moral, as opposed to a practical, education. There were also regional differences. The northern states, particularly in New England, tended to pursue more progressive educational policies, with towns being required to establish schools funded, in part at least, by local taxes. In the south, most schools were private ventures, with the more affluent parents hiring itinerant tutors to home-school their children. The late-eighteenth-century introduction of grammar schools and academies may have incorporated arithmetic into the school curriculum, but the standard of attainment seems to have been very low.

In colonial times, arithmetic had been seen as a somewhat "vulgar" skill of use only to merchants and those involved in commerce, confined mainly to the major towns and cities. To the majority of the population, situated in the vast rural expanse of the country, the study of arithmetic was considered to be of no use whatsoever. But with the growth of a market economy and the incursion of commerce into everyday life, the early years of the American republic saw a re-thinking of educational priorities. Arithmetic, it was argued, could not only foster commercial acumen but also the mathematical rationality needed for the burgeoning new democracy. The 1840 census reported that roughly 55 percent of school-age children were attending elementary schools. The provision of tax-subsidized elementary schools in every state by 1870 naturally increased this figure, although such schools were far more plentiful in the towns than in the countryside. By 1900 school attendance was compulsory in 34 out of the 45 states, with the school-leaving age having been raised to 14 or higher in thirty of them. Finally, in 1918, the completion of an elementary school education became compulsory for every child in the United States.

Although girls had access to an elementary education throughout this period, in both Britain and America such access was far from universal and was always inferior in quantity and quality to that provided to boys. Female education was regarded as a mere afterthought to the more important task of teaching their male counterparts (Harris 1997). While boys received instruction in reading, writing and arithmetic, in girls' education, tuition in the latter was frequently replaced or supplemented by classes in needlework, a skill that, while it required some basic numeracy, was regarded as more appropriate for the subservient role expected of them. Not only would women have no need for numerical skills in later life, it was argued, but it was strongly believed that the female brain was incapable of the sustained effort that mathematical thought required. And although later in the period, young women could also be exposed to activities which required a modicum of arithmetical knowledge—such as cookery, which can be seen as requiring a basic understanding of proportion, and household management, which entailed living within a budget—for the most part it was considered that to teach women anything higher than the most basic numerical skills was a waste of time.

For racial minorities in America, the situation was even worse. Native Americans, in addition to enduring decades of coercion and removal from ancestral lands, were also the recipients of a number of misguided attempts to educate their children in the Western manner. Residential schools were established, mainly run by religious organizations, which Native American children were forced to attend; and this policy persisted well into the twentieth century. As far as African Americans were concerned, before the Emancipation Proclamation of 1863, it was illegal even to attempt the education of a black slave. After the Civil War of 1861–5, publicly funded schools for black children were subject to habitual underfunding and chronic overcrowding. Virginia was fairly typical among the Southern states in having a relatively low overall percentage of African-American children in its schools. Yet, despite this, in 1900, the average black school in Virginia still had 37 percent more pupils than the average school for white children. By the outbreak of the First World War, the proportion of African Americans receiving an elementary education was still well below the national average.

The influence of educational practices in both Britain and America can be seen far beyond their borders in the style of arithmetical education imparted to children in several non-Western countries during this period. In China, for example, after a series of unsuccessful military confrontations with Britain, France, Russia, and Japan, indigenous mathematical and scientific methods were increasingly rejected in favor of Western techniques in an attempt to modernize and strengthen the country. British and American missionaries led the way in establishing a Western presence, via the establishment of educational institutions, libraries, and the translation of textbooks on arithmetic (and other subjects) into Chinese. French and German educational approaches also had an impact. By the beginning of the twentieth century, firm foundations had been laid for the Chinese to begin to educate their own pupils in Western mathematics and science.

For much of the nineteenth century, Japan was even less receptive to foreign influence than China had once been. For over two centuries from 1639, during the Edo period (1600–1868), Japan was effectively closed to outsiders. This self-imposed isolation resulted in the development of a distinctive form of largely computational Japanese mathematics, known as "wasan." The provision of education in reading, writing, and wasan across the country was quite widespread, even in remote villages, with almost every town having private elementary schools ("terakoya") teaching both boys and girls. Despite the isolation, certain aspects of Western knowledge managed to seep in, especially after the signing of the treaty of amity with the United States in 1854. The first Western-style arithmetic textbooks appeared in 1857, with the full-scale adoption of Western methods coming soon after the beginning of the Meiji era in 1868. From 1872, only Western mathematics was permitted to be taught in schools, with the adoption of Arabic numerals and Western notation. By the end of the Meiji period in 1912, Japan had been transformed, with elementary mathematics education being just one facet of a society that had changed beyond recognition in less than half a century.

Perhaps the most striking example of the merging of Western educational practices with an indigenous mathematical tradition was in India. Elementary education already existed in a variety of elementary schools ("pathshalas" for Hindu children and "madrasas" for Muslims), although this was primarily for upper-caste boys. The homegrown Indian style of learning arithmetic placed a much higher emphasis on mental computation; as a British contemporary reported, "the power which the little schoolboys display in mental arithmetic is quite astonishing to the European" (Senthil Babu 2012: 39). As in China, British involvement in Indian education began informally via missionaries, but after around 1813 the East India Company became increasingly involved in the provision of education and from the 1830s, with an increase in funding from the British government, there was a decisive shift towards imparting European knowledge to Indian pupils. British arithmetic textbooks were either adapted for Indian use or fully translated, and standard techniques of blind memorization and repetitive exercises were imposed. By the 1850s, pupils in the pathshalas were being taught to perform calculations in the British style, on slate tablets, and in the more traditional form of mental arithmetic, or "bazaar mathematics." As well as establishing their own schools, attempts were also made by the British authorities to convert existing schools into Western-style institutions; but the pathshalas survived well into the twentieth century. And despite official recommendations in both 1854 and 1882 to increase the availability of elementary education to both lower castes and girls, efforts to do so met with limited success.

TEXTBOOKS

At this elementary level, the mathematical instruction was largely arithmetical in content, but a modern student would no doubt be surprised by the relative size and scope of the discipline in comparison to what is learnt today. This is best illustrated by a representative survey of some of the arithmetic textbooks published during the course of the century (Denniss 2009). They feature a wide array of theoretical and practical topics, from basic operations, fractions and proportion to interest calculations, conversions of weights and measures, and commercial transactions. Although often intended for the instruction of school children or young people, these books were often written in language more accessible to teachers and adult learners.

For this reason, and since textbooks could also be costly, it was relatively unusual for pupils to actually own copies themselves. Instead, those who could afford to do so would often fill elaborate exercise books with worked examples and problems as a permanent record of their education and possibly for future reference. These copybooks, neatly written and properly bound, were as much exercises in calligraphy and decorative penmanship as they were in mathematics. The opening page often featured an elaborate title, with the student's name, the date and sometimes additional embellishments from the teacher. This practice of copying out worked examples as a keepsake, which seems to have died out around 1900, existed both in Britain and America during the eighteenth and nineteenth centuries, and many examples have been preserved.

In one such specimen from *c.*1825, we see a certain Abraham Lincoln, then aged 16, working on the following problem: "[I]f 100£ in 12 months gain 7£ interest what principal will gain 3£-18-9d in 9 months." As can be seen in Figure 1.2, after converting 3 pounds, 18 shillings and 9 pence into 945 pennies, dividing by the product of 7 and 9, multiplying by 100 and 12, and converting the result back into pounds, the young Mr. Lincoln has deduced the correct answer of 75 pounds.

The emphasis on rote-learning and copying, combined with the absence of a sound understanding is evident in many manuscript notebooks. Worked examples often follow definitions copied laboriously either from a blackboard

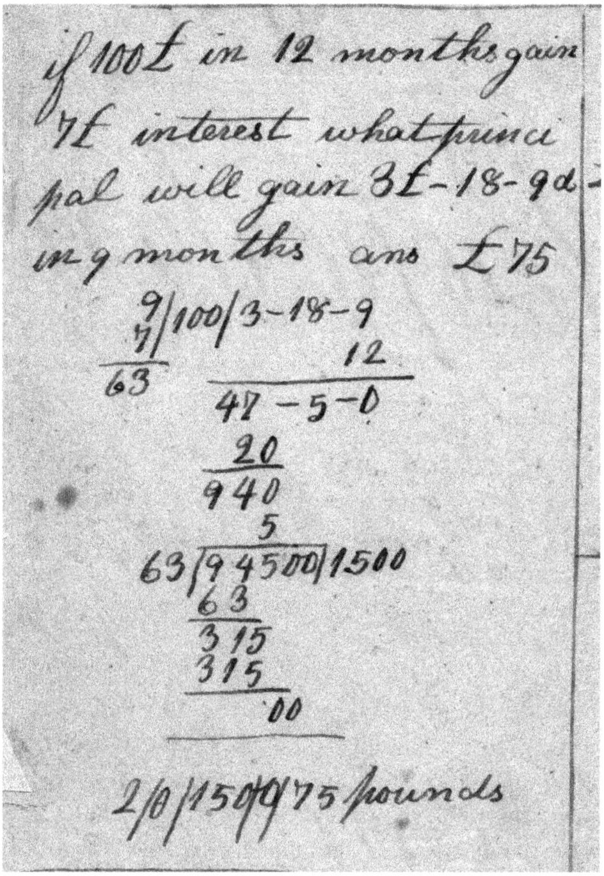

FIGURE 1.2. Part of a leaf from Lincoln's mathematical exercise book, *c.* 1825. MS Am 1326, Houghton Library, Harvard University. By permission of Houghton Library.

or textbook via dictation, such as this monstrosity written by an 11-year-old boy in 1817:

> Inverse proportion is when more requires less, and less requires more. More requires less is when the third term is greater than the first, and requires the fourth term to be less than the second. And less requires more is when the third term is less than the first, and requires the fourth term to be greater than the second.[1]

Having memorized definitions and rules, students would have been expected to apply them to problems, churning out solutions with as little independent thought as possible. Prolix rules, inefficient procedures and illogical (or nonexistent) justifications abound, indicative of the overall quality of contemporary instruction in the subject. Copybooks such as these provide firm evidence that, at least in the early part of the nineteenth century, skill in arithmetic was believed to be best acquired via memory rather than understanding.

Throughout the first half of the nineteenth century, the best-selling and most widely used arithmetical book in Britain was, without doubt, Francis Walkingame's *The Tutor's Assistant*. First published in 1751, it sold thousands of copies, ran through scores of editions and remained in print until at least 1885. The natural question, when perusing a copy today, is why? As its title suggests, it was clearly written for the teacher and not the student, and indeed without a tutor's guidance it would have been completely useless to the pupil. Its explanations are so obscure and perfunctory as to be virtually nonsensical, as witness the copybook definition of inverse proportion above, which comes, word for word, from page 44 of the 33rd edition of 1806. Walkingame's definition of subtraction on page 4 is similarly useless: "This being the Reverse of Addition, you must borrow here (if it require) what you stopped at there, always remembering to pay it to the next." But nobody expected quality of exposition in an arithmetic textbook, so in this respect, Walkingame's text differed little from his contemporaries. The key to its success lay not in the ability to elucidate the subject matter, but was rather due to the hundreds of examples and exercises (with answers) that the book provided. Plus, it was cheaper than most of its competitors.

The increasing number of schools in operation, and the inclusion of arithmetic in the curriculum, soon led to criticisms of existing treatments of the subject. Writing in 1812, Thomas Clark maintained that no work on arithmetic existed in English that adequately introduced the four arithmetical operations with sufficient clarity and rationale to satisfy the beginner. An attempt had been made in John Bonnycastle's *Introduction to Arithmetic* in 1810, a learned but dry work that presented the material in a coherent and ordered fashion, with each subject featuring a clearly worded definition, rule, illustration, proof, and further examples and exercises. Some of these, while no doubt well intentioned, had quite unrealistic expectations with regard to accuracy, which the author noticeably failed to meet with his answer in the following case:

> From 400 years, take 98 years 3 mo. 8 hr. 10 sec.
>
> *Ans.* 301 years 9 mo. 3 wk. 6 dys. 15 hrs. 59′, 50″.

In any case, at 400 pages, its cost would have been prohibitive for all but the wealthiest students.

The first commercially successful challenge to the dominant style of textbook came with the publication of *Elements of Arithmetic* by Augustus De Morgan.

This work, expanded over six editions between 1830 and 1876, introduced a new style of exposition designed to be read by both teacher and pupil—although no doubt teachers would have been the main beneficiaries. In addition to increased clarity and precision, De Morgan presented the material in a new, more user-friendly way, beginning with simple motivating examples before introducing rules, demonstrations, and more sophisticated problems. He also encouraged the use of pebbles, counters, or other simple objects to enable the beginner to better conceptualize numerical quantities.

This use of tangible objects was a new and innovative approach to the teaching of basic arithmetic and shows the definite influence of the contemporary Swiss educational reformer Johann Heinrich Pestalozzi, who famously advocated environments in which children could learn through activities and exploration, rather than the more traditional approaches. This philosophy was the driving force behind Horace Grant's *Arithmetic for Schools and Families*, published in several editions from 1841. An exceptionally clear work, this book was designed to be read to children between the ages of eight and twelve by the teacher (or parent). It features lucid explanations with plentiful examples, and although the topics covered are fairly standard, there is also a strong emphasis on the practice of the newly defined skill of "mental arithmetic."

But these liberal ideas were never as influential in Britain as they became in the United States, where such methodologies were more readily adopted. Indeed, America in the 1820s saw a huge rise in the popularity of mental arithmetic, inspired by the work of its Harvard-trained popularizer, Warren Colburn. His *First Lessons, or Intellectual Arithmetic on the Plan of Pestalozzi* (1821) and *Arithmetic Upon the Inductive Method of Instruction* (1826) were both hugely influential and ran to multiple editions. Colburn's goal was to abandon the dependence on rules and rote-learning, and to foster logical thinking and understanding. The skill of mental arithmetic was imparted, beginning with the use of tangible objects, via spoken words to abstract numbers. In this way, children were encouraged to do simple calculations in their heads without recourse to rules, memorization or even writing. The idea was that through this technique of learning via exploration, children would discover the basic rules of arithmetic via inductive reasoning, making the mental leap from particular cases to general laws. In reality, the number of children who were capable of discovering mathematical laws unaided was always going to be small. Nevertheless, by virtue of their massive sales and substantial influence, Colburn's books were tremendously successful in ushering in a new pedagogical philosophy in the United States.

Yet, despite Colburn's success, serious barriers remained to the teaching of numeracy and numerical reasoning. Possibly the most troublesome of all such arithmetical obstacles was the dreaded "Rule of Three." Stated in its simplest form, problems of this kind required the student, given the values of three known quantities, to find a fourth. They usually proceeded along the lines of, "If the ratio of quantity *a* to *b* is equal to the ratio of quantity *c* to a mystery quantity, what is this mystery quantity?" Stated in algebraic terms, the question becomes almost trivial because, if

$$\frac{a}{b}=\frac{c}{x}$$

then clearly $x=bc/a$. But in the nineteenth century numerical questions involving proportion were considered arithmetical rather than algebraic, particularly since algebraic reasoning was considered far too advanced for most people. Thus, solving problems involving the rule of three became contingent on memorizing verbose and confusing rules, such as those from Walkingame, and attempting to apply them.

The rule of three usually involved solving contrived applied arithmetical problems such as currency exchange, partnerships and wills, discount and interest, the filling and emptying of wells, tare and tret (namely, making allowances for the weight of containers and spillage of contents in shipping), objects moving at different velocities, and the completion of manual tasks by laborers working at different speeds. For example: "How long would it take 15 men to build a wall if 45 men can do it in 10 days?" The more traditional books would instruct the student to set up the ratio 15 / 45 = 10, multiply the second and third numbers together (giving a nonsensical product of 45 men × 10 days), and then divide by the first to obtain the correct answer. However, newer authors like Colburn, Grant and De Morgan were now providing far more intelligible explanations: since 45 men would take ten days, it would clearly take one man 45 × 10, or 450 days to do the same thing. Therefore fifteen men would complete the job in a fifteenth of that time, or in other words, 30 days.

Still, for every such enlightened author offering intelligible explanations, there were just as many old-style teachers writing dull texts crammed with mechanical procedures, not all of which were up to the task of solving the problems they posed. One such conundrum was contained in Frederick Emerson's *North American Arithmetic*. Its third volume of 1834 featured the following problem of compound proportion on page 286:

> If 12 oxen eat up $3\frac{1}{2}$ acres of grass in 4 weeks, and 21 oxen eat up 10 acres in 9 weeks, how many oxen will eat up 24 acres in 18 weeks; the grass being at first equal on every acre, and growing uniformly?

This question, known as the "pasturage problem," quickly became the most celebrated mathematical problem in mid-nineteenth-century America, following a pledge by the National Teachers' Association in June 1835 to award the considerable sum of $50 for the most "lucid analytical solution." It took over three years to award the prize (to a teacher from Boston), and nearly a quarter of a century before the correct solution, $37\frac{113}{175}$, was finally published in 1859. In that time, the prize committee had received only 48 correct solutions out of 112 submissions—all from schoolteachers and none from students, for whom the problem had originally been intended.

The fact that the question featured a misprint was only discovered in the 1870s, when it was revealed that the problem had originated in Isaac Newton's *Arithmetica Universalis*, where it was stated that: "12 oxen eat up $3\frac{1}{3}$ acres of grass in 4 weeks." This correctly worded problem produced a much more realistic answer of 36 oxen and had also featured (without attribution) in Walkingame's *Tutor's Assistant*. It had subsequently been commandeered (again without attribution, but with the misprint) by Emerson, adding further confusion to an already difficult problem, by forcing its solution to introduce fractional quantities, thus making "mincemeat" of the oxen! The whole episode illustrates not only an extreme case of computational complexity in a textbook problem, but also the limitations of purely arithmetical procedures to provide a satisfactory solution.

Back in Britain, the influence of more progressive authors such as Grant and De Morgan was beginning to show. For example, the Scottish mathematician Edward Sang, in his *Elementary Arithmetic* of 1856, placed a far greater emphasis on fostering computational skill than many authors, with the importance of the use of decimals, as well as vulgar fractions, being stressed. Sang's book was also significant in being the first on the subject to relegate the rule of three to a mere application of a more general treatment of ratio and proportion. Meanwhile, John Brook-Smith's *Arithmetic in Theory and Practice*, in print for over forty years from 1860, aimed explicitly to foster clarity of understanding and a firm grasp of principles. His definition and explanation of the rule of three extended to over seven clearly written pages and, like Sang's treatment, it followed from a discussion of ratio and proportion. Brook-Smith illustrated the method with several fully worked examples, such as: "If 7 men can mow a field in $18\frac{3}{4}$ hours: in how many hours can 15 men mow the same field?" The comparative clarity of his solutions stood in marked contrast to the obscurity of many earlier and contemporaneous authors:

> If the number of men be *increased*, the number of hours in which they can mow the field will be proportionally *decreased*, hence the *inverse* ratio of 7 men to 15 men equals the ratio of $18\frac{3}{4}$ hours to hours required; that is,
>
> 15 men : 7 men = $18\frac{3}{4}$ hours : hours required;
>
> $\therefore$ hours required $= \frac{75}{4} \times \frac{7}{15} = \frac{35}{4} = 8\frac{3}{4}$ hours.[2]

By the time of the first publication of Charles Pendlebury's highly popular *Arithmetic* in 1886, the rule of three was no longer even mentioned by name.

With the educational reforms of the nineteenth century came the increasing dominance of an exam culture in both British and American society, and this is particularly reflected in arithmetical textbooks of the later part of this period. The textbooks by Brook-Smith and Pendlebury contained substantial collections of sample examination papers, with hundreds of questions (and answers) from entrance examinations for schools, universities, the military, and the civil service. During the first half of the century, the prevailing convention had been to include the answer at the end of each question, as in the example from Bonnycastle above. But as the century wore on, the answers began to be relegated to the back, or in many cases omitted altogether, being published separately in dedicated answer keys. This model proved extremely successful, with Pendlebury's *Arithmetic* in all its various versions—*New School Arithmetic*, *A Shilling Arithmetic*, *Commercial Arithmetic*, and so on—requiring new editions almost every year for at least two decades, in print runs of 20,000 copies per edition, and remaining in print until after the Second World War. But by then, textbooks dedicated to inculcating basic numeracy were becoming rarer. And with the educational and technological changes of the twentieth century, the need for a textbook devoted solely to arithmetic would be eliminated entirely.

WEIGHTS AND MEASURES

The difficulty of much of the material in nineteenth-century British and American arithmetic textbooks was further compounded by the need to include "real-world" numerical problems of the kind that could be of potential utility to their readership. Such problems frequently involved financial, commercial, and other practical considerations, including computations of weight, volume, area, speed, and other attributes. But these questions were rendered especially complicated by a particular cultural idiosyncrasy that distinguished Britain (and its colonies) and the United States in this period—namely their shared, but distinct, systems of weights and measures.

The British system of weights and measures had evolved over many centuries from a variety of Roman, Saxon, and medieval units of measurement. Many now standard units such as the pound, gallon and mile originally had significantly different definitions and uses, often varying by locality. For example, at 1976 yards, the Scottish mile was 12.3 percent longer than the statutory English mile. Throughout the late medieval and early modern period, various attempts at standardization were introduced, until by the seventeenth century, several units of measurement had received definitions that more or less conform to their current values. Nevertheless, numerous differences remained, as the definition of certain units depended on what was being measured. There were, for example, three separate definitions of a gallon (for wine, beer, and corn, i.e., dry measure), while the most well-known distinction concerned troy and avoirdupois weight, with the former being used primarily for medicine, precious metals and stones, and the latter for all other goods.

Regional variations also persisted. A survey of 1693 found that "a pound of butter might weigh just 16 ounces at Huntingdon but 36 ounces in Cheshire. A stone was just 8 pounds at Hertford and Hitchin, 16 at Leeds, Halifax and Dorchester, and 18 pounds in many northern counties. And a bushel was 8 gallons at Brecon, Pembroke and many other places, but as much as 32 for wheat and rye at Chester, even 40 for oats at the same place" (Hoppit 1993: 83). Meanwhile, a plethora of irregular and non-standard measures were also in use across the regions. A British government report in 1820 listed over 150 different local units of weight and measure, including the bag, bale, basket, bottle, box, bucket, bunch, cabot, cart, cartload, cord, dish, drop, ell, faggot, flask, garb, gunny, head, heap, hoop, hutch, jar, jug, keel, kiver, knot, leap, mark, mast, measure, merk, mug, pack, packet, piece, pig, pot, sack, spindle, stick, teal, tankard, thread, truss, tub, waggon-load, web, and windle.

To formally codify the nebulous practices across Britain and her growing empire, the Weights and Measures Act of 1824 established a unified "Imperial" system based on a limited number of standard units, and abolished many, but by no means all, of the lesser-known measurements. All measures of length were derived from the standard Imperial yard, all weight was based on the troy pound, and the three separate gallons were replaced by a single gallon for all liquids. This new measure was defined to be such that a gallon of distilled water at 62°F weighed exactly 10 pounds, which, since a gallon contained eight pints, resulted in the well-known saying:

> A pint of pure water
> Weighs a pound and a quarter.

But this standardization did little to obviate the fundamental computational complexity inherent in the system. In length for example, there were 12 inches in a foot, 3 feet in a yard, 5½ yards in a rod, 4 rods in a chain, 10 chains in a furlong, and 8 furlongs in a mile. But how, other than numerical skill or simple blind memorization, could one know that a mile comprised 1760 yards?

The Imperial system came into effect in 1826; but since Britain's former American colonies were no longer under British rule, none of these changes applied to the United States. Indeed, the Americans preserved many of the old definitions abandoned by the British. As a result, for example, a hundredweight remained 100 pounds in America, but changed to 112 pounds in Britain, resulting in the British ton (or 20 hundredweight) weighing 12 percent more than its American counterpart. This change also explains why the US pint of

16 ounces is substantially less than the 20 ounce British pint: although in both countries, a pint is one-eighth of a gallon, the United States retained the old British definition of a gallon of wine, which was roughly 80 percent of the volume of the new Imperial gallon.

Not surprisingly, merchants or businessmen involved in trade between the two nations (or indeed, any nations) would necessarily have to be able to perform conversions between the two systems of units, or else employ someone who could. Moreover, with such bewildering subdivisions of measurements as grains, drams, and scruples, it was just as important to be able to convert units within one's own system. It was therefore far from uncommon to find problems of the following kind in British and American nineteenth-century textbooks:

> If 25 labourers can dig a ditch 220 yards long 3 ft. 4 in. wide and 2 ft. 6 in. deep in 32 days when the day is 9 hours long, how many labourers would be able to dig a ditch half a mile long 2 ft. 4 in. deep and 3 ft. 6 in. wide in 36 days when the day is 8 hours long?[3]

This gives some idea of the computational burden that the use of Imperial, or "English," measurements entailed. Yet, this paled into insignificance when combined with the additional complexity that arose from the British monetary system.

MONEY

As with its convoluted system of weights and measures, Britain's currency had a long history stretching back over a period of more than a millennium. Its fundamental unit had been in use since at least the eighth century when, following the lead of Charlemagne in mainland Europe, it was divided into 20 shillings (corresponding to the French *sous*, derived from the Latin *solidus*), which in turn comprised 12 pennies (corresponding to the French *deniers*, derived from the Latin *denarius*). This gave a total of 240 pennies—which, in the days when pennies were made of silver, weighed one troy pound, thus giving the fundamental unit of currency its name. With the word pound deriving from the Latin *pondus*, meaning weight, and its symbol (£) being an ornate letter L for *libra*, the Roman word for balance, the abbreviations £, s., d. for pounds, shilling, and pence had been in use for centuries. Thus, anyone brought up in nineteenth-century Britain would have been familiar with this notation, expressing 3⅝ pounds, for example, not as £3 150d., nor 72s. 6d., nor even 870d., but as £3 12s. 6d.

In addition to requiring a working knowledge of modular arithmetic, the situation was compounded further by various other denominations—such as the farthing, crown, and guinea—all of which combined to produce a computationally challenging monetary system:

1 penny = 4 farthings
1 groat = 4 pence
1 shilling = 12 pence
1 crown = 5 shillings
1 pound = 4 crowns = 20 shillings
1 guinea = 21 shillings

Given that the average annual income of the majority of workers in Victorian Britain was unlikely to exceed £40, to be financially numerate, people needed to be conversant, or at least moderately familiar with, these various subdivisions. The effective implementation of wage payments, household management, medical bills, investments and insurance all relied on the ability to compute with the monetary units correctly. A baker wishing to buy 3 hundredweight of sugar priced at 4¾d. per pound would need to be able to budget for the expense. A draper who bought 150 yards of cloth at 14s 6½d. per yard would need to know how much to charge if he wanted to sell it all at a profit of £39 13s. 9d. And a farmer with annual overheads of £389 19s. 8d. in wages, £68 16s. 5½d. in rates, £39 5s. 8d. in taxes, and £404 4s. 7¾d. in livestock would have been well advised to be able to work out his yearly expenditure. But this was by no means a simple matter.

Things were more straightforward on the other side of the Atlantic, at least in theory. As one of the first nations to institute a decimal system of currency in 1792, the United States exhibited a somewhat progressive attitude to the complexity of financial calculations expected of its citizens. It was Thomas Jefferson who had proposed the new system—10 milles to a cent, 10 cents to a dime, and 10 dimes to a dollar—with the relative ease of computation being one of its main justifications:

> The facility which this would introduce into the vulgar arithmetic would, unquestionably, be soon and sensibly felt by the whole mass of people, who would thereby be enabled to compute for themselves whatever they should have occasion to buy, to sell, or measure, which the present complicated and difficult ratios place beyond their computation for the most part. (Cline Cohen 1999: 128)

But, as others were quick to stress, the new currency also played a political role, by helping to democratize the acquisition of knowledge in the new republic. In the words of early nineteenth-century congressman Erastus Root:

> let us, I beg of you, Fellow-Citizens, no longer meanly follow the British intricate mode of reckoning. – Let them have their own way – and us, ours. – Their mode is suited to the genius of their government, for it seems to be

> the policy of tyrants, to keep their accounts in as intricate, and perplexing a method as possible; that the smaller number of their subjects may be able to estimate their enormous impositions and exactions. But Republican money ought to be simple, and adapted to the meanest capacity. (Cline Cohen 1999: 129)

Yet, despite his sanguinity that, by virtue of its simplicity, the new federal currency would soon drive out all competitors, in the 1811 edition of his *Introduction to Arithmetic*, Root admitted that he still had to give many of the examples in pounds, shillings and pence, observing pragmatically that: "The customs of a great nation cannot be wholly changed in a month nor a year." However, British denominations were far from the only competitors to federal money in the early days of the US Mint, as a variety of foreign currencies were accepted as legal tender throughout the United States in the first half of the nineteenth century. Chief among them was the Spanish dollar, which accounted for roughly a quarter of all transactions as late as 1830. It was not until 1857, shortly before the American Civil War, that Congress ultimately ended the status of foreign coins as legal tender and Jefferson's goal of simplifying financial arithmetic across the United States was finally achieved.

Roughly contemporary with this, a more influential and far more thoroughgoing reform was underway in revolutionary France, which was to transform both currency and weights and measures. In 1790, the French national assembly proposed an international system of measurements having as its basis a unit of length based on the earth's circumference. Intended to simplify computations for the general public, this system was to be decimal, with larger and smaller multiples of each unit arrived at by dividing or multiplying by powers of 10—but France's proposal was rejected by the British Parliament and the United States Congress. In response, the French developed their own "metric" system unilaterally, formally adopting it at the end of 1799, with implementation beginning in 1801.

The fundamental unit, defined as being one ten-millionth of the distance between the North Pole and the Equator through Paris, was called the "meter," from the Greek word for "measure." All other measurements of length (such as the centimeter, millimeter, kilometer) were derived from it, as were other measurements, such as the gram—the mass of one cubic centimeter of water. Meanwhile, the old French currency of livres, sous and deniers was replaced by a decimal currency of francs, decimes and centimes, not dissimilar to that introduced in America. However, because of poor implementation, the new metric system proved unpopular, resulting in its partial suspension by Napoleon in 1812. Its reintroduction in 1840 was more successful, and by the late 1850s, its use in France had become firmly entrenched. By this time, via trade and colonization, it had also begun to spread across Europe and other parts of the world, until by 1914 over 70 countries and territories had formally adopted the metric system.

All the same, conservative Britain and her growing empire held out steadfastly in refusing to adopt any kind of decimal system. In the immediate aftermath of the defeat of Napoleon, any measures to promote decimalization, no matter how well intentioned, smacked of revolution and republicanism. A proposal brought by the British MP John Wrottesley to decimalize the British currency was firmly rejected by Parliament in 1824, but attempts to promote change continued sporadically throughout the century. A lobby group called the Decimal Association was founded in 1841, which at its height boasted 1,500 members, including prominent mathematical scientists of the day, such as Charles Babbage, John Herschel and Augustus De Morgan. Not surprisingly, their main arguments focused on the relative computational simplicity of a decimal system. Their immediate aim was for the institution of a decimal currency, to be followed quickly by a decimal system of weights and measures.

In response to growing pressure, the government launched a new denomination of British coin in 1849: the florin. This silver coin, worth two shillings, precisely one-tenth of a pound, was conceived as an experimental first step towards a possible decimalization of the coinage. A change to a decimal currency began to look increasingly likely, to the extent that sections on decimal coinage began to appear in British arithmetic textbooks. Following the Great Exhibition of 1851, which had highlighted the use of a decimal system for international trade, a parliamentary committee was appointed to consider the question in 1853. Among those to give evidence was Thomson Hankey, a recent Governor of the Bank of England, who stated that, not only did he use the decimal system in private calculations, but that he believed the introduction of a decimal currency would lead to an improvement in education, and tangible benefits for trade, industry, and all sections of the population.

It was not difficult to see his reasoning. An American consumer, knowing that an item was priced at $253.75 per hundredweight, would be able to deduce almost instantly that, since an American hundredweight was 100 pounds, this item would cost $2.5375 or $2.54 per pound. But a British customer, faced with an equivalent price of £253 15s. per hundredweight (or 112 pounds) would first have to divide £253 by 112, to obtain £2, with a remainder of £29. This remainder would then have to be multiplied by 20, and a further 15 added to obtain the remaining number of shillings, 595. This would then have to be divided by 112, giving 5 shillings, with a remainder of 35 shillings, or 420 pence. Finally, dividing this 420 by 112 would give $3\frac{3}{4}$ pence. Only then could the final cost of the item be deduced: £2 5s. $3\frac{3}{4}$d. per pound.

Hankey proposed that the pound, or "sovereign," should be retained as the principal unit of currency, but divided into 1000 "mils" with the following new coins:

half-sovereign (500 mils)
florin (100 mils)
half-florin or shilling (50 mils)
quarter-florin or "Victoria" (25 mils)
10 mils or "groat"
3 mils
2 mils
1 mil

This proposal, however, like many others, foundered under the objection that the general populace would be unable to convert quantities of the old and new currency, simply because for certain amounts no easy conversion was practicable:

> The great drawback to the proposed system is that no sum of money of the present coinage, except a multiple of 6d, can be paid *exactly* in the new coinage, for 1d is $4\frac{1}{6}$ mil; hence a fare, a toll, or a stamp of 1d, must be paid for either by 4 mil or 5 mil, that is either by $\frac{1}{6}$ mil too little or by $\frac{5}{6}$ mil too much; or 2d either by 8 mil or 9 mil, that is either by $\frac{1}{3}$ mil too little or $\frac{2}{3}$ mil too much, etc.; and therefore in the *transition* from the present to the proposed system there must be considerable inconvenience.[4]

Other contemporary proposals included reducing the value of the farthing slightly, so that there would be 1,000 farthings to the pound instead of 960, after which new coins could be issued: 10 farthings to a cent, 10 cents to a florin, and 10 florins to a pound. But although a Royal Commission on Decimal Coinage commented favorably on the idea of decimalization in the mid-1850s, the then Chancellor of the Exchequer, Sir George Cornewall Lewis, argued that the time was not yet right for a change on the grounds of the multiplicity of proposals before him. This prompted the following sarcastic response from Augustus De Morgan:

> . . . we beg permission to add two [proposals] of our own invention . . . Our first plan is 10 farthings a what's-his-name; 10 what's-his-names a how-d'ye-call-it; 10 how-d'ye-call-it's a thingembob. Our second plan is 10 farthings a George; 10 Georges a Cornewall; 10 Cornewalls a Lewis.[5]

Sarcasm aside, there were those who went further, perhaps too far, in their zeal for decimalization. In addition to money, weights and measures, the author of *The Decimal System as a Whole*, published in 1856, urged for the decimalization of everything, even clocks and calendars. De Morgan reviewed this book in his usual pithy manner:

FIGURE 1.3. Augustus De Morgan (1806–71). Photo by Hulton Archive / Getty Images.

> The proposition is to make everything decimal. The day, now 24 hours, is to be made 10 hours. The year is to have ten months, Unusber, Duober, etc. Fortunately there are ten commandments, so there will be neither addition to, nor deduction from, the moral law. But the twelve apostles! Even rejecting Judas, there is a whole apostle of difficulty. These points the author does not touch.[6]

Politicians nevertheless remained cautious: non-decimal modes of reckoning were too firmly entrenched in British society, the expense of a change would be too great, and the challenge of educating the general public was too daunting. When the Royal Commission presented its final report in 1859, it concluded that, as long as the British system of weights and measures remained non-decimal, no change should be made to the currency.

This was how things would remain for over a century. Despite various revivals of the debate in subsequent decades—and even the legalization of metric units for all purposes in 1897 (they had been legal in the United States since 1866)—the use of non-decimal units in weights, measures and money remained firmly in place in Britain and her empire well into the twentieth century. Indeed, as international competition increased in the run-up to the First World War, opponents of decimalization maintained that the computational difficulty inherent in Britain's traditional measures was a decisive argument in favor of their retention, since this numerical complexity put British traders at an advantage in the international market. As one journalist wrote candidly in 1911: "we enjoy an advantage over foreigners because we can readily grasp their metric measures while they will never be able to understand ours" (Kula 1986: 282). It would not be until 1971 that Britain would finally see the introduction of a decimal currency.

AIDS TO COMPUTATION

The fact that arguments in favor of decimalization were largely unsuccessful throughout the nineteenth century arguably reveals the extent to which certain numerical systems and computational techniques were ingrained into the cultural fiber of both British and American societies. But it might also be explained in part by the fact that, then as now, very few people actually undertook difficult numerical calculations entirely unaided, and so did not feel that an easier system of computation was needed. Just as today, the mere fact that someone was taught arithmetical procedures in their youth was no guarantee that person would retain the knowledge. And as we have seen, there were still many to whom schooling in arithmetic was unavailable. For these reasons, it was fortunate that computational assistance was readily available in two forms: the slide rule and the ready reckoner.

The slide rule was an invention of the mid-seventeenth century that in its most basic form resembled a simple ruler with a set of sliding scales, enabling the user to perform quick and easy multiplication and division. More sophisticated versions allowed more advanced operations such as finding powers, roots, exponentials, logarithms and trigonometrical values. Although relatively rare and expensive at the beginning of the nineteenth century, the increased manufacturing capabilities of the period soon enabled cheaper mass production of these analog devices, until by 1850 De Morgan could write that: "for a few shillings, most persons might put into their pockets some hundred times as much power of calculation as they have in their heads." By the latter part of our period, as a result of improvements in design and affordability, their use had become widespread.

While slide rules were useful for general numerical calculations, ready reckoners became indispensable tools for computations involving specific units of measurement. Unlike slide rules, they were simply printed collections of tables, but like the slide rule, they could also trace their origins to the mid-seventeenth century, when they were published with titles like *A Platform for Purchasers* and *Accompts of Merchandise Ready Computed.* It was after the publication of Daniel Fenning's *Ready Reckoner, or Trader's Most Useful Assistant* in 1757 that the name "ready reckoner" began to become dominant among such compendia. A common aid to calculation throughout the Western world from the eighteenth century, they were particularly widespread in Britain and America, due obviously to the complexity of the multi-base weights and measures in use—although it should be noted that the usage of ready reckoners dropped off substantially in the United States in the early nineteenth century, as the decimal currency became more widespread.

With the rise of low-cost publishing from the mid-nineteenth century, these cheaply printed pocket books became near ubiquitous commodity items in Britain. They contained page after page of tables showing multiples of unit costs for commodities by quantity, weight, or volume, with prices ranging from farthings to hundreds of pounds. By the late nineteenth century, they had become almost essential in commercial transactions. The baker unsure of his arithmetic, for example, could now discover in a matter of seconds that 3 hundredweight of sugar at 4¾d. per pound would cost him £6 13s. But ready reckoners were not just for calculating purchases—they contained a variety of numerical tables for a host of everyday purposes, including interest computation, the conversions of weights and measures, tax calculations, currency exchange, and calendrical reckoning. And due to their immense utility, their use, along with the equally convenient slide rule, continued right up until the 1970s, when the introduction of a decimal currency and the rise of affordable pocket calculators rendered them obsolete almost overnight.

ACCESS TO NUMERICAL DATA

Ready reckoners were far from being the only source of numerical data available to the working population. In both Britain and America, the nineteenth century saw the growth and widespread publication of compendia, libraries of "useful knowledge," and other works of reference for the self-improvement of the working population. Enlightened bodies such as the Society for the Diffusion of Useful Knowledge in Britain—and its identically named American counterpart in Boston—produced tables, encyclopedia, and scientific dictionaries, not only to educate the working and middle classes but also to provide the appropriate computational tools with which professionals such as builders, architects, carpenters, surveyors, salesmen, shopkeepers, mechanics, or engineers could practice their trades effectively. Especially popular in both countries were almanacs: annual collections of quantitative facts and statistics. Of particular use to farmers and those engaged in outdoor work, almanacs featured numerous calendars, astronomical and meteorological data, along with information on sunrise and sunset times, phases of the moon, eclipses, tides, postal rates, customs and excise duty, and copious other numerical facts concerning agriculture, trade, industry, and demographics. Information on the latter was updated via numbers garnered from the growing resource of governmental data collection: the decennial census.

The United States Census had been mandated by the Constitution of 1789, which stipulated that, for purposes of taxation and representation, the population must be counted every ten years. Before the creation of the Census Bureau, the first six censuses (1790–1840) were conducted by US marshals, who hired assistant marshals to conduct the actual enumeration. Over time, and as the country grew, the range of the data also expanded. Initially, only the heads of household were recorded by name, along with a demographic count of the remaining household members. From 1850, all names were included, with full information on all household members, including slaves. Attempts were also made, from 1860, to include Native Americans, although early results were far from comprehensive. By mid-century, the number of questions had increased to incorporate data on manufacturing, taxation, vagrancy and crime. In fact, so great was the volume of information obtained that it took eight years to publish all of the data from the 1880 census.

In Britain, the first nationwide census took place in 1801, motivated by a concern that the size of the population might outstrip the country's resources, particularly with regard to food. Although at first little more than a basic headcount, by 1841 the surveys were becoming more sophisticated. This was the first year in which the census was administered by the General Register Office, formed in 1837 to record and report on demographic trends. Under the GRO, the census questions expanded to include names of every household member, along with age, sex, occupation, and place of birth. By 1851, forms required information as to whether any household member was blind, deaf or dumb, with the categories "imbecile," "idiot," and "lunatic" being added in 1871. Although the number of questions asked was consistently lower than for the US census, by 1911 the questions had expanded to include information on mortality of children, duration of marriage, and the industry or service by which the head of household was employed.

In the United States, the first attempt to count Americans who were "insane" or "idiotic" occurred in the sixth census of 1840. The results proved controversial when data published from the census seemed to reveal that disturbingly high numbers of black people resident in northern non-slave-owning states suffered from mental illness, in marked contrast to the correspondingly low figures for the Southern states, where the majority of black people were slaves. Naturally, those in favor of slavery immediately proclaimed the results as providing strong evidence of the damaging effects of slave emancipation, with the consequent beneficial effects of maintaining the status quo that this implied. Equally unsurprisingly, abolitionists insisted that the data on the published returns must have contained errors. Subsequent analysis revealed this to be correct, with errors later found to have occurred at both the data-entry and at the computational level. It marked a turning point in the history of numeracy in America, when it was realized for the first time, not only that numbers do not always present an objective truth, but that they can be manipulated and misrepresented just as easily as words.

While the debacle of the 1840 census highlighted limitations in the arithmetical abilities of US civil servants, it ultimately led to more stringent measures to control the accuracy of the data produced. Nevertheless, for much of this period, the tabulation of numerical information relied fundamentally on the (fallible) basic arithmetical skills of human "computers." In Britain, the GRO supplemented this manpower with the purchase of a number of mechanical "arithnometers" between 1870 and 1903, and by 1895 could boast that the accuracy of all calculations was being checked once with an arithnometer and once with a slide rule. Meanwhile in the United States, an electromechanical tabulating machine was invented by Herman Hollerith specifically for use in the census of 1890 (see Figure 1.4). Data was entered on punched cards and tabulated by machine, thus reducing the processing time considerably. This combination of human and machine-powered number processing thus allowed both the British and United States governments to present previously overwhelming amounts of data in formats that could be intelligible to the general public.

The explosion in the accessibility of numerical data took place within the wider context of the Victorian culture of data collection. Nineteenth-century scientists, economists, and politicians all valued the information that large amounts of data would give them. Numerical data was particularly important, with demographic data in particular leading to various social and public health reforms, as well as improved actuarial life tables. The increased availability of cheap, affordable printing gave the public greater access to numerical data, of which ever-increasing amounts were published by the decade. By the opening years of the twentieth century, more numerical information was available at less cost to more people than at any previous time in history.

CONCLUSION

The nineteenth century witnessed a huge rise in the visibility of numbers and the perceived uses of numeracy in Anglo-American society. The expansion of education on both sides of the Atlantic meant that by 1914 basic tuition in the rudiments of arithmetic was available to a far wider proportion of the population than in 1800. Higher literacy rates plus the availability of vast numbers of affordable books on the subject also meant that self-improvement in numerical

FIGURE 1.4. Hollerith tabulating machines in use during the 1890 US Census. *Scientific American*, 1890.

skills was readily available to motivated individuals. And, despite official reluctance to simplify the abstruse system of weights and measures, healthy sales of ready reckoners and other numerical aids in the latter half of the period show that resources to ease the computational burden were extensively used. Yet, for all the advances that were made with regard to everyday numeracy, the substantial developments in trade, commerce, transportation, industry and technology during this period also introduced new computational challenges, unknown to previous generations.

Time calculation, for example, had never been much of an issue before the Industrial Revolution, because pinpoint accuracy was never required. For the most part, simply knowing the approximate time of day was all that was needed. But the Victorian era introduced a previously unknown complication to the arithmetic of timekeeping. The expansion of transport and communication during the nineteenth century created a need for a unified timekeeping system. Before the introduction of railways in Britain, the fact that noon in London occurred ten minutes before noon in Bristol had never been a problem. Nevertheless, by the 1830s, it had become a very serious one, as inconsistencies in timekeeping were causing an increasing number of near misses and even some train crashes. In 1840, the Great Western Railway controversially introduced "Railway Time," by imposing London time across the whole network, the first occasion that time had been synchronized between different locations to a single standard.

In America, the same problem required a more ambitious solution. With the westward expansion of the United States throughout the century, and each city operating a different time standard, operators of the new railway lines needed a plan that would offer a uniform train schedule for departures and arrivals. The result was the introduction of the four standard time zones for the continental United States in 1883. The following year, again to facilitate calculations involved in transport, communication, and commerce, international time zones based on the use of Greenwich Mean Time were adopted at the International Meridian Conference in Washington, DC. (It should be noted, however, that the international systematization of timekeeping did not eliminate all errors, as evidenced at the 1908 London Olympics when the Russian shooting team arrived twelve days late due to their country's adherence to the Julian calendar!)

So, was the higher profile of numbers and numerical reasoning reflected in a greater overall numerical ability among the general population? This is not an easy question to answer, as information is piecemeal. However, it does not appear that the increase in exposure to numerical reckoning during this period necessarily led to a uniform increase in numerical skill. The fiasco of the 1840 US census provides a perfect illustration of the woeful innumeracy of civil servants and the general public mid-century. And even by the end of the period,

FIGURE 1.5. This cartoon by Joseph Swain from an 1887 edition of *Punch* calls for a more pragmatic form of education for the laboring classes in Victorian Britain. Photo by The Cartoon Collector / Print Collector / Getty Images.

the introduction of machines to assist in the calculations and tabulations of the various censuses is a telling sign, not of the clerks' increased numerical skill, but rather of the authorities' lack of confidence in it. Meanwhile, the public's use of repositories of numerical data such as almanacs required no arithmetical skill whatsoever, while the huge popularity of ready reckoners to perform any conceivable everyday calculation is indicative of the pervasive lack of numerical skill among the general population. Finally, even though an elementary education in arithmetic was, in theory, available to all, in practice, large sections of the population such as females and minorities were grossly underserved, and even for those privileged enough to receive a thorough grounding, by the end of the period instruction had progressed little further than the rote learning of rudimentary reckoning skills.

The general lack of skill and ability was also coupled with an inherent suspicion and mistrust of the use of numbers. While some, like the Belgian social scientist Adolphe Quetelet and the British social reformer Florence Nightingale, extolled the new subject of statistics as a tool for social improvement and reform, others like Charles Dickens regarded the use of numerical data with deep suspicion, even hostility. The 1854 novel *Hard Times* was a satire against the mid-Victorian use of numbers and statistics, or as Dickens called it, "the wickedest and most enormous vice of this time." Objecting in particular to measures that viewed society on the basis of averages, Dickens deplored the use of cold facts and figures to obstruct, or justify the lack of, progressive legislation. In his eyes, and those of similarly minded Victorians, the blind use of numbers, averages, and statistics would never result in a better society. This general skepticism of the use of number in social and political matters was summarized effectively in words attributed to Benjamin Disraeli by Mark Twain, and no doubt captured the mood of many: "There are three kinds of lies: lies, damned lies, and statistics."

Thus we see that, although the basic skills that comprised everyday numeracy stayed relatively constant throughout the nineteenth century, what changed over time were the potential applications of those skills. Whereas in 1800 the principal uses of basic arithmetic had been largely confined to matters concerning commerce and household management, by 1914 its use had extended to include the realms of data collection, finance, politics, travel, social reform, epidemiology, public health, and mass education. The need for basic numeracy had never been greater and the expansion of its influence was increasing. This trend would continue throughout the twentieth century, as more fields of human activity came to rely on the analysis of numerical data, bringing more areas of modern culture into the realm of numbers, and vastly increasing the applicability of numeracy to everyday life.

CHAPTER TWO

Practice and Profession

National Models for Professionalization

TOM ARCHIBALD AND DAVID E. ROWE

Professionalization in mathematics, as we know it today, largely evolved over the course of the nineteenth century, a development brought about by institutions that not only promoted research but also trained a new generation of researchers. Innovation in admissions and in teaching methods were hallmarks of the newly founded École polytechnique in Paris, an elite school for training engineers, who had to master large doses of mathematics. The social status of higher mathematics had already taken a dramatic turn upward over the course of the French Enlightenment, whose proponents helped mold the political ideals of the Revolution. Mathematics, as taught at the École polytechnique, thus came to serve as a gateway to the higher echelons of the French meritocracy (Bottazzini and Dahan-Dalmedico 2001). Under Napoleon, who cultivated close personal relations with leading mathematicians, in particular Gaspard Monge and Joseph Fourier, France embraced the spirit of a new society dominated by technocratic elites. This message, reflecting the significance of mathematics as a harbinger of social change, echoed throughout Europe during the first decades of the century. A very different educational model emerged, however, in Prussia, an eastern German state that acquired territories along the Lower Rhine in 1815.

Napoleon had earlier abolished the tiny feudal states in this westernmost part of the Holy Roman Empire, which also ceased to exist after 1806. The new regimes set up under French rule as the Confederation of the Rhine introduced numerous legal and institutional reforms. A defeated Prussia fell under French occupation, yet Wilhelm von Humboldt nevertheless pursued plans for a new university in Berlin in 1809 (renamed Humboldt University, in honor of Wilhelm and his brother Alexander von Humboldt, after the Second World War). Berlin University spearheaded an educational movement in Prussia, which eventually led to widespread reforms at other universities during the period that culminated with the political unification of the German states in 1871 (Begehr et al. 1998). This educational system aimed not so much to foster the acquisition of technical knowledge and skills, but rather to promote spiritual development, or *Bildung*, mainly by pursuing neo-humanist ideals. According to this conception, there was no better way to build character and discipline in young men than by immersing them in the world of classical antiquity. At the *Gymnasien*, which were elite secondary schools, Greek and Latin dominated the curriculum; mathematics was valued as well, but mainly as a tool for developing the mind. Within the Anglo-American educational tradition, a gymnasium was associated with sports, in particular gymnastics (*Turnen* in German). In Germany, on the other hand, the gymnastics movement (*Turnbewegung*), founded by Friedrich Ludwig Jahn in 1807, began outside the educational system and was a major factor in spawning German nationalism. Sports only entered the curricula of German schools in the latter half of the century and then with special emphasis on gymnastics and physical fitness.

Within the German university system, the larger task of *Bildung* fell to the philosophical faculties, which emerged from the former arts faculties of the medieval universities. The French essentially secularized and demoted their universities, while creating a system of *grandes écoles* focused on mathematics, science, and engineering. In the German states, on the other hand, the universities gradually reformed by upgrading the status of their philosophical faculty vis-à-vis the three higher faculties—theology, medicine, or law—that prepared students to enter the corresponding professions. To obtain a doctoral degree in philosophy (a PhD) it no longer sufficed to demonstrate scholarly competence alone: faculties began to demand original research in the form of a written thesis supervised by an expert in a special field. This required expanding the faculties so that they could cover all disciplines in the natural sciences and humanities. At the same time, their graduates now had a clear professional objective: most would go on to become teachers at the *Gymnasien*. Those who qualified to teach at the upper level of these prestigious schools, the *Oberlehrer*, occupied a rung in German society only slightly below the exalted position of the university professors. While in retrospect this might appear as a liberalizing trend, participation remained limited to men from wealthier families until late in the century, and (in Prussia for example) the guiding hand of the ministry was present in every professorial appointment and policy development.

This trend traces back to 1737 and the founding of Göttingen University in Hanover, but by the early nineteenth century it became part of the reform movement implemented by the Prussian state. Thus, whereas the French created a new kind of elite, comprised of scientist-engineers, the Germans placed humanists, especially philologists and philosophers, at the pinnacle of a quite different type of intellectual elite. For mathematics, both of these competing models enjoyed impressive success and each exerted a strong influence on institutions in other nations, both within Europe and beyond, as part of a general modernization process. Out of this international mix emerged a kind of template for what we recognize today as the professional mathematician.

Professionalization also went hand in hand with growing specialization, as witnessed by a proliferation of journals devoted exclusively to the publication of research in pure and applied mathematics. This situation contrasts sharply with the role played by professional mathematicians during the early modern period (see Chapter 2 in Volume 3), which covered mathematical practices ranging from conducting ballistics tests to composing horoscopes. The locus of activities in higher mathematics shifted during the eighteenth century to the learned academies in Paris, Berlin, and St. Petersburg. The many leading mathematicians who worked at these institutions often undertook projects of a more mundane nature, whereas celestial mechanics was one of their dominant theoretical interests. By the end of the nineteenth century, the notion of a natural philosopher in the tradition of Isaac Newton had disappeared giving way to a new breed: the natural scientist. The new term aptly fits figures like Hermann von Helmholtz or Lord Kelvin (Smith and Wise 1989), but it no longer applied to Henri Poincaré or David Hilbert; even if their interests were exceptionally broad: they were mathematicians. By 1900, the trend toward specialization thus led to the formation of separate national societies for the disciplines of mathematics, astronomy, and physics. At the same time, these trends led to the growth of a class of professional mathematicians, paid to do and to disseminate mathematical research while imparting their ideas to others (for philosophical implications, see Ferreirós 2016).

In surveying these developments, we first consider important social and institutional innovations that took place in France, the dominant mathematical culture in Continental Europe up until 1848. Its École polytechnique served as a kind of model for the polytechnic schools founded in various German states (Karlsruhe, Munich, and Dresden) as well as in Zurich and in the Italian states, notably Piedmont. French textbooks were also quite popular in Germany, largely because professors at the universities lectured without using a text and often on material that went beyond this literature. By mid-century, mathematicians at leading German universities had begun to spawn various schools of mathematics with a strong accent on pure research in analysis, geometry, algebra, and number theory. A latent rivalry between France and Germany turned particularly bitter in the wake of the Franco-Prussian War of 1870–71. With the founding of the German Empire, which closely followed the unification of Italy, nationalism became a potent ideology that strongly colored the period 1870–1914. We therefore have taken the year 1870 as a convenient dividing line for describing how the practices and professions of mathematicians changed over the course of the long nineteenth century.

MATHEMATICAL TRADITIONS IN MODERN STATES, 1800–1870

In the wake of the French Revolution, Continental mathematics underwent rapid transformation, sparked by profound political and social changes that reconfigured European science as well as its institutions of higher education (Grattan-Guinness1990). In France, Enlightenment ideals of social progress animated a series of educational reforms, which unleashed an unprecedented explosion of scientific activity in Paris during the Napoleonic period. Mathematicians like Monge, Fourier, Lazare Carnot, and even the aged J.-L. Lagrange played a prominent part throughout, Carnot serving on the Directoire, the group of 5 men who were heads of state. Mathematicians occupied an important place within the French scientific community, both as researchers as well as educators. Carnot and Monge were instrumental in the founding of the École polytechnique. During the eighteenth century, these had been largely distinct activities, but now academicians found themselves thrust into a new role: to train the nation's technocratic elite, a task which set a premium on their ability to convey mathematical ideas clearly. Mathematics thus had a role equivalent to that of legal scholarship related to the Napoleonic code for the wider bureaucracy.

Lazare Carnot was less influential than Monge mathematically, yet every bit as important as a public figure. In fact, aside from Napoleon himself, probably no single actor contributed more toward saving the French Republic than Carnot, who as a member of the Committee of Public Safety took charge of the military. In 1794, acting as Minister of War, he re-organized the French Revolutionary Army following authorization of a *levée en masse* by the National Convention. France was then facing a desperate situation trying to fend off the combined strength of armed forces from five nations. The Rhineland and Belgium had already fallen, the Austrians were invading Flanders, and the Prussians were attacking in the Alsace. Meanwhile British forces were occupying Corsica and Toulon, as their fleet threatened to take control of the Mediterranean Sea.

The *levée en masse* nearly tripled the number of French troops; some were patriots burning to fight, others ordinary conscripts. Carnot placed these new recruits under experienced soldiers to ensure they received proper training and discipline. He also devised new military tactics that soon turned the tides of war in France's favor. Rather than confronting enemy forces head on, he divided the massive French army into small units that could mobilize quickly, outmaneuvering the foe while conducting flank attacks. Carnot retreated from political life after Napoleon crowned himself emperor in 1804, but he later accepted the title of Count of the Empire, and during the Hundred Days he served as Minister of the Interior. With the fall of Napoleon, Carnot fled into exile to evade charges of regicide because of his role in the execution of King Louis XVI. He died in Magdeburg, Prussia, but the Third Republic restored his honor in 1889 by interring his remains in the Panthéon.

FIGURE 2.1. Statue of Gaspard Monge by François Rude in Musée Rude, Dijon. Creative Commons: Photography by Christophe Finot.

Monge had been Minister of the Navy during the Revolution and later joined the entourage of scientists who took part in Napoleon's expedition to Egypt. Like Lagrange, Monge was a member of the Napoleonic Senate (*Sénat conservateur*), over which he presided for one year. After his beloved emperor fell from power in 1815, Monge lost all his honors and offices, dying in disgrace. Pierre-Simon Laplace, on the other hand, managed to find favor with each passing regime, gathering titles throughout his illustrious career. In 1806, Napoleon made him a count of the Empire, whereas under the Bourbon Monarchy he became a marquis in 1817. Most fitting of all was the informal title he enjoyed as the "French Newton," a reputation he gained with his five-volume *Mécanique Céleste* (1799–1825), which aimed to demonstrate the stability of the solar system. Newton had thought that universal gravitation alone could not account for this and that a stable universe required divine intervention. Laplace consciously dispensed with this hypothesis.

On the other side of the political spectrum stood the pious Catholic, Augustin-Louis Cauchy (Belhoste 1991). A professor at the École polytechnique and staunch Bourbon loyalist, Cauchy found the post-1830 regime of Louis-Philippe so distasteful that he felt compelled to leave France rather than take an oath to serve it. As a mathematician, Cauchy's only peer was Carl Friedrich Gauss; both excelled in pure and applied research, though Cauchy was the

FIGURE 2.2. Entrance of the École polytechnique, rue Descartes. Lithograph by Charles Kreutzberger (1829–1909), illustration from Le Musee Francais, n 42, June 1858. Photo by DEA / Biblioteca Ambrosiana / Getty Images.

more prolific author with some eight hundred articles and five textbooks. Among these was his influential *Cours d'analyse* (1821), which he wrote for his course at the École polytechnique. In it, Cauchy stressed the importance of rigor in analysis by replacing the heuristic notion of infinitesimals (generally used in teaching) with something close to the concept of limits developed and taught by Karl Weierstrass in Berlin. This passage from rough rules of thumb to strict definitions, while unpopular with students, soon became an important tool for the practicing research mathematician (see Chapter 3). Cauchy's *Cours d'analyse* was the first in a long series of French textbooks on the calculus that bore the same title, replacing older titles emphasizing the calculus of differentials and integrals. Indeed, French textbooks were to set the standard throughout the century, especially in Germany.

The national mathematical community in France was large enough to support specialized journals in mathematics from the Napoleonic period onwards. The *Journal de l'École polytechnique*, reserved for those with connections to that institution, was established with the founding of the school, and like the Academy's journals could count on funding from the state. The *Annales de Mathématiques Pures et Appliquées*, published in Nîmes from 1810 to 1832 by Joseph Diaz Gergonne, featured an idiosyncratic mix focusing on geometry and analysis, bringing innovative pure-mathematical work from all over France to the public eye. Gergonne also published foreign work, notably by the German geometers Jakob Steiner and Julius Plücker. Mathematical articles also appeared in the *Bulletin de Férussac*, a Paris-based reviewing journal with broad coverage.

French scientific leadership also made Paris attractive for students from outside France and for visiting researchers. While the French system made it difficult for foreigners to complete degrees, lectures at the Collège de France were open to male auditors, who could apply for permission to attend lectures elsewhere in some circumstances. Thus, the Germans P. G. Lejeune Dirichlet and Plücker studied in Paris in the 1820s, when Dirichlet in particular came under the influence of Fourier. Dirichlet's work in number theory attracted the attention of leading French mathematicians already in the 1820s, including Fourier and Siméon-Denis Poisson. The educational and social systems could also support the unusual case of Sophie Germain, who attended lectures by proxy and whose bourgeois family status and connections allowed her contacts with leading mathematical figures.

Until the 1830s, the Paris Academy of Science remained practically a world unto itself, an exclusivity that often proved discouraging for budding talents. The two most dramatic cases of this involved Évariste Galois and the Norwegian Niels Henrik Abel, both talents of the first order who died in their twenties, though in fact their treatment was more or less typical. Decades later, older contemporaries gradually came to grasp the profundity of their work, which deeply influenced research in algebra and function theory during the latter half of the century. A more open attitude toward "outsiders" emerged, however, when Joseph Liouville arrived on the Parisian scene. His journal, founded in 1836, provided a voice for mathematicians outside the Paris Academy and even to some outside France. Liouville's young protégé, Charles Hermite, along with the aged A.-M. Legendre, understood the importance of the new theory of elliptic functions and higher transcendentals cofounded in the 1820s by Abel and C. G. J. Jacobi. Thus, by mid-century research trends in the pace-setting mathematical communities of France and Germany showed a renewed emphasis on fields located on the pure end of the mathematical spectrum.

The employment situation in France, however, was rather bleak, especially since the universities, which were closed during the revolution, remained vestigial until a revival in the 1870s. The importance of research for practicing teachers, while downplayed during the restoration, received a boost in the 1850s through the work of Louis Pasteur at the École normale supérieure, a revolutionary foundation devoted to teacher education. Pasteur's own work on public health and vaccination exemplified the social importance of research, and he courted students with the idea that ENS was a place to learn to do research. This seduced the young Gaston Darboux, who in 1860 placed first in the entrance examinations for both the EP and the ENS, and startled the nation by choosing the ENS. A decade later, mathematicians had largely hijacked Pasteur's journal, the *Annales scientifiques de l'ENS*.

Whereas Paris completely dominated the mathematical sciences in France, the situation in Germany was by contrast highly decentralized. The employment situation was in general better for mathematicians than in France, partly due to a proliferation of military and technical schools, but mainly because a doctoral degree came to be a normal qualification for those who taught the upper-level courses in the *Gymnasien*. Nevertheless, it was at the universities that mathematics came to thrive. Beginning in the 1830s, three universities in German states sparked the beginnings of what soon became a vibrant research tradition in mathematics. These were Königsberg, where Jacobi taught until 1844, the newly founded university in Berlin, where Jacobi joined Dirichlet and Steiner, and Göttingen, where Gauss had long taught until his death in 1855. It was, however, Gauss' successor, Dirichlet, who truly inaugurated Göttingen's tradition as a center for mathematical research (Merzbach 2018).

If Dirichlet's decision to study in France was exceptional, his autodidactic outlook certainly was not. Gauss, too, was a largely self-taught mathematician, as were Dirichlet's contemporaries, Jacobi and especially Steiner. E. E. Kummer and Weierstrass taught for several years in *Gymnasien* before gaining university positions. Other prominent mathematicians, like Hermann Grassmann and Hermann Schubert, taught in schools throughout their entire careers. None of these outstanding figures had an academic mentor in the modern sense. Indeed, what Jacobi and Dirichlet managed to achieve through their teaching in Königsberg, Berlin, and then Göttingen was to plant the seeds for the kind of mentoring that became commonplace at the German universities during the latter half of the century. This transformation was at once subtle and profound: subtle because it took place informally through the formation of mathematical schools with distinguished leaders; its profundity is reflected in today's culture whenever mathematicians talk about their mentors, or by looking up someone's academic lineage on the Mathematics Genealogy Project, a data base with well over 200,000 names. For Dirichlet, Jacobi, and their followers, mathematics was no longer just a body of knowledge and special techniques; it was a calling, something one professed, much like a religious faith. Indeed, the German word for profession is *Beruf*, from the verb *rufen*, meaning to call. Those who followed the leader of a mathematical school often did so with a sense of reverence and discipleship. These features of mathematical schools were by no means unique; very similar characteristics prevailed in other disciplines as well. They formed a large part of the ethos that promoted a new research imperative (Turner 1981), which played such a key role during the formative period when modern research universities emerged in Germany.

The Prussian universities spearheaded this movement, just at the time that the Prussian monarchy seized the initiative for unifying the German states by defeating the Austrians on the battlefields of 1866. Bismarck attained his goal of uniting Germany under Prussian hegemony five years later with the defeat of the French. Italian unification was a far more difficult affair, in part owing to Austrian control of parts of northern Italy as set forth by the Congress of Vienna in 1815. The period thereafter, known as the *Risorgimento*, culminated in 1871 with the founding of the Kingdom of Italy. The *Risorgimento* was a broad-based cultural, political and social movement that promoted Italian unification by appealing to romantic, nationalist and patriotic ideals as a kind of renewal of the Italian renaissance.

Three mathematicians—Francesco Brioschi, Enrico Betti, and Luigi Cremona—played an active part in this movement. All three were involved in the failed military insurrection of 1848 that attempted to end Austrian rule in Italy. After unification, each served in the new Italian government, either as senators or in the Ministry of Education. They were educational reformers who helped to develop research centers at universities. Cremona came to Rome in 1873 to organize the Royal College of Engineering. Two of his works from this time appeared in English translations: *Graphical Statics* and *Elements of Projective Geometry*. Italian mathematics reflected the influence of French institutions and ideas, while at the same time drawing on recent trends in Germany. In 1858, Brioschi, Betti and Felice Casorati visited several major mathematical centers in France and Germany. This trip helped to acquaint them not only with research trends in these countries but also gave them the chance to establish personal ties with several leading mathematicians. Betti met and befriended Bernhard Riemann in Göttingen, and he later invited him to Pisa in 1863. Riemann made three trips to Italy and died there in 1866. During this time, Betti was director of the Scuola Normale Superiore in Pisa. This elite institution, founded during the Napoleonic era, served as a counterpart to the École normale supérieure in Paris.

As suggested earlier, professionalism in mathematics gained considerable visibility over the course of the nineteenth century through journals that specialized in publishing mathematical research. Many of these journals, which mainly served national interests, exist to the present day. A. L. Crelle's *Journal für die reine und angewandte Mathematik,* dating from 1826, is still known today simply as "Crelle." In 1836, Liouville founded the *Journal de mathématiques pures et appliquées,* and the first important Italian journal, *Annali di scienze matematiche e fisiche,* was founded by Barnaba Tortolini in 1850. Eight years later, physics no longer appeared in the title, which was renamed *Annali di matematica pura e applicata.* By the 1870s, under the

FIGURE 2.3. Since 1847 the Scuola Normale Superiore in Pisa has been located in the Palazzo della Carovana. Photo by Education Images / Universal Images Group / Getty Images.

primary editorship of Brioschi, but assisted by Cremona, Betti, Casorati, and Eugenio Beltrami, the *Annali* had become one of the leading international journals.

A similar development took place in Great Britain (Flood, Rice, Wilson 2011), where the first research periodical devoted to mathematics appeared in 1836: the *Cambridge Mathematical Journal.* After only four volumes, however, the title changed to *The Cambridge and Dublin Mathematical Journal,* reflecting the prominence of these two centers for research. After only nine more volumes appeared, this name was dropped in 1855 in favor of *The Quarterly Journal of Pure and Applied Mathematics*, which was mainly edited by James Glaisher up until his death. Glaisher was a student and later Fellow of Trinity College, Cambridge, where he was second wrangler in 1871; he also edited the *Messenger of Mathematics.* Since Newton's time, Cambridge had enjoyed special prestige as the elite school for training those who sought careers in the mathematical sciences (Howson 1982). The Mathematical Tripos was the most prestigious examination in Britain and competition to succeed was fierce. Several of those who were most successful—like three second "wranglers," G. G. Stokes, William Thomson (later Lord Kelvin), and James Clerk Maxwell—would later dominate British mathematical physics in the second half of the century, at the same time as they brought mathematics to bear on significant instruments of technological change, including telegraphy and electricity (see Chapters 5 and 6). The Tripos also served as a springboard for mathematicians who went to work in the colonies, where they established mathematical outposts and observatories, notably in India and Japan, the latter with the support of the Meiji government after its 1868 decision to "Westernize."

The outstanding American mathematician during the first half of the nineteenth century was Nathaniel Bowditch, a prominent figure in Boston's scientific circles remembered for his English translation of Laplace's *Mechanique celeste.* During his work on this massive project, he engaged young Benjamin Peirce, who was long associated with Lawrence Scientific School at Harvard University. There Peirce attracted several outstanding pupils, including Simon Newcomb and George William Hill. This suggests a line of interest in celestial mechanics that ran from Bowditch to Hill, the latter a leading theoretical astronomer who did groundbreaking work on the three-body problem. Hill introduced the differential equation that now bears his name in conjunction with his studies of lunar motion. This equation is equivalent to an infinite system of algebraic equations, which Hill solved by calculating the associated determinant. Afterward, under Newcomb's supervision at the Nautical Almanac, Hill took up work on the mutual perturbations of Jupiter and Saturn. His prolific output led to the four large volumes that comprise *The Collected Mathematical Works of George William Hill.*

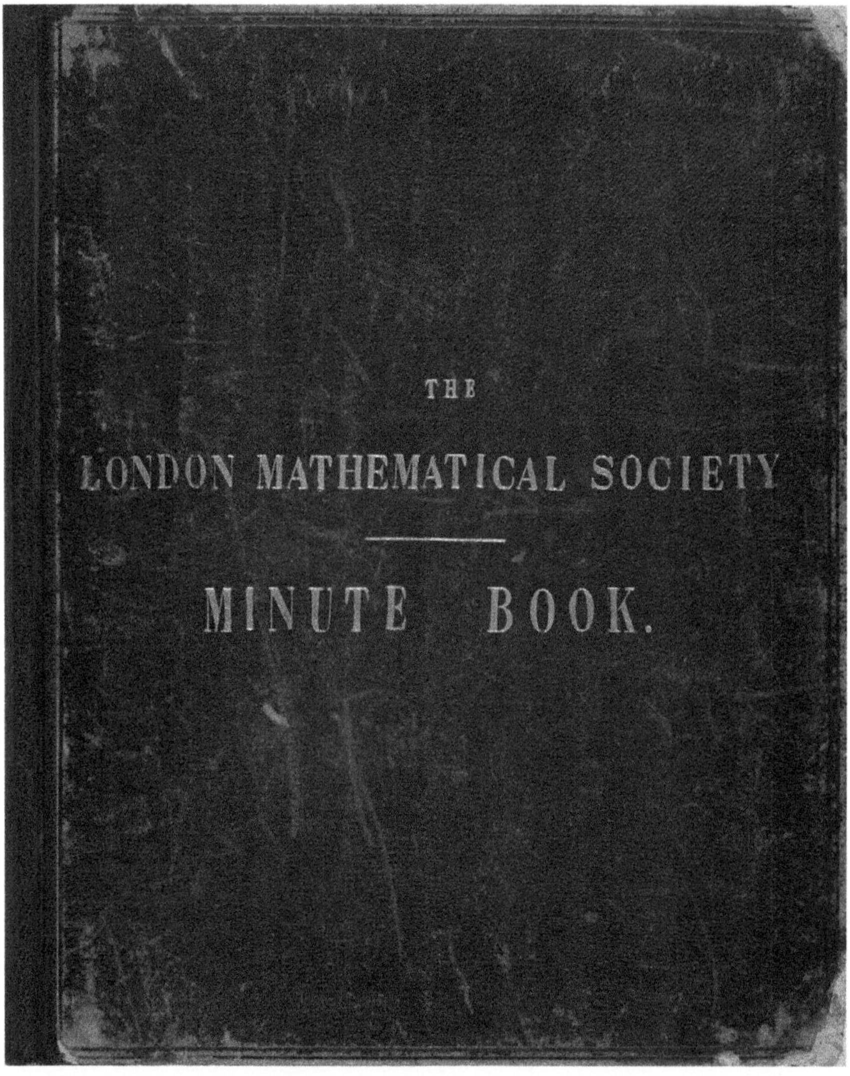

a

FIGURE 2.4a–b. The founding members of the London Mathematical Society, 1865. Courtesy of the London Mathematical Society.

FORMATION OF NATIONAL COMMUNITIES AND SOCIETIES, 1870–1900

The trends toward professionalization and specialization sharply accelerated during the period after 1870, augmented in part by a substantial growth in the number of research mathematicians. By 1900, independent mathematical research communities with their own organizations had formed in Britain,

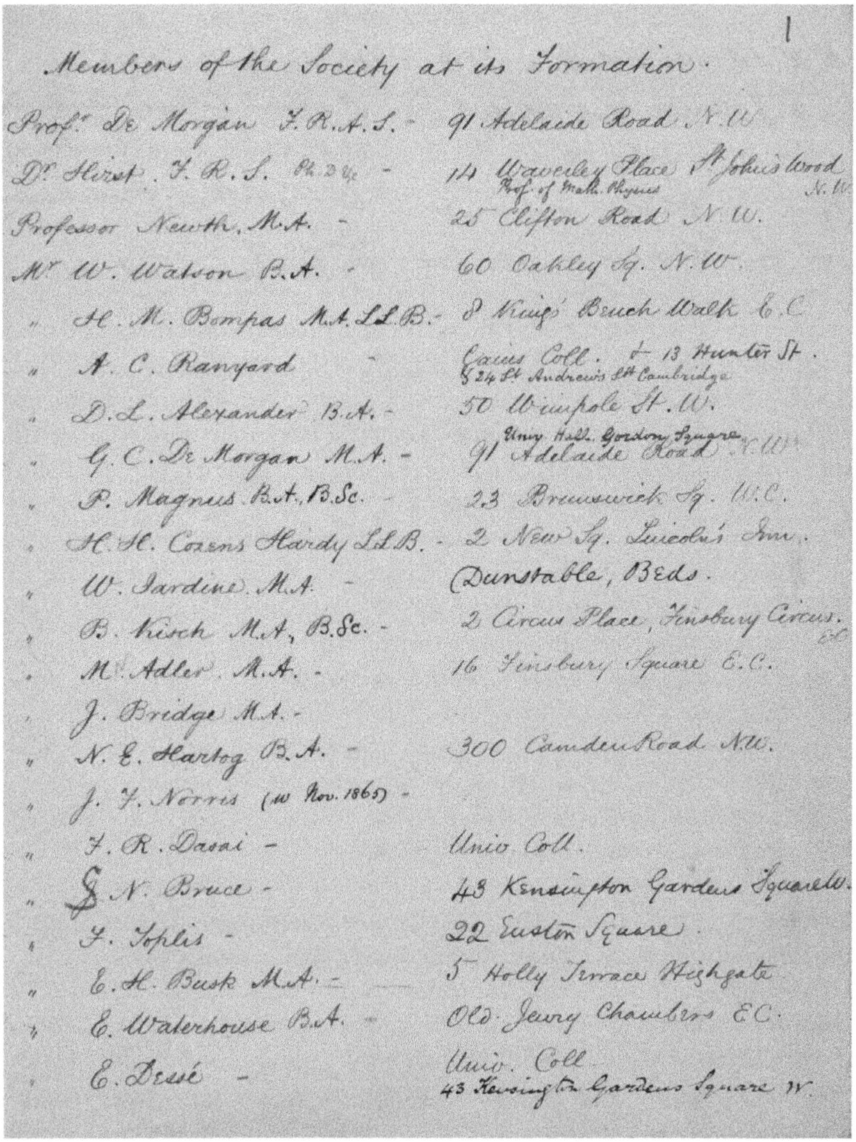

1

Members of the Society at its Formation.

Profr De Morgan F.R.A.S. -	91 Adelaide Road N.W.
Dr Hirst. F.R.S. Ph.D. -	14 Waverley Place St John's Wood N.W. Prof. of Math. Physics
Professor Newth, M.A. -	25 Clifton Road N.W.
Mr W. Watson B.A. -	60 Oakley Sq. N.W.
" H. M. Bompas M.A. LL.B. -	8 King's Bench Walk E.C.
" A. C. Ranyard -	Caius Coll. & 13 Hunter St. & 24 St Andrew's St Cambridge
" D. L. Alexander B.A. -	50 Wimpole St. W.
" G. C. De Morgan M.A. -	Univ. Hall. Gordon Square 91 Adelaide Road N.W.
" P. Magnus B.A., B.Sc. -	23 Brunswick Sq. W.C.
" H. H. Cozens Hardy LL.B. -	2 New Sq. Lincoln's Inn.
" W. Jardine M.A. -	Dunstable, Beds.
" B. Kisch M.A., B.Sc. -	2 Circus Place, Finsbury Circus. E.C.
" M. Adler M.A. -	16 Finsbury Square E.C.
" J. Bridge M.A. -	
" N. E. Hartog B.A. -	300 Camden Road N.W.
" J. F. Norris (w Nov. 1865) -	
" F. R. Dasai -	Univ. Coll.
" S. N. Bruce -	43 Kensington Gardens Square W.
" F. Toplis -	22 Euston Square.
" E. H. Busk M.A. -	5 Holly Terrace Highgate
" E. Waterhouse B.A. -	Old Jewry Chambers E.C.
" E. Desse -	Univ. Coll. 43 Kensington Gardens Square W.

b

France, Germany, Italy, and the United States; soon thereafter, research schools and mathematical societies would emerge in Russia, Poland, Sweden, and other countries. Through visits and by attending conferences and international congresses, members of these communities and local centers began to intensify their contacts, building new networks of power and influence. These developments gradually led to a transformation in conventional modes of production and communication among mathematicians, a shift shaped by a complex variety of factors—technical, social, political, educational, etc.—that affected nearly all forms of scientific endeavor in various ways. Traditional links with subjects like astronomy, geodesy, and mechanics survived, though by 1900 these ties had loosened due to rapidly diverging professional research interests. At the same time, the social status and function of mathematicians, as producers and purveyors of mathematical knowledge, underwent significant changes.

The period following France's defeat in the Franco-Prussian War turned out to be a fertile one for mathematics in France, in part due to institutional reforms spurred by the defeat. The government of the Third Republic took the view that the defeat was due in part to moral weakness, for which an improved educational system would be a remedy. Over the period from 1871 to 1914, curricula and training methods came under intense scrutiny. This in turn led to a considerable expansion of employment opportunities for well-trained mathematicians. There had been little incentive under the Second Empire (1851–70) to obtain a doctorate, for example, since the degree merely enabled its holder to pursue a very small number of geographically dispersed post-secondary teaching positions. Indeed, France produced only one or two mathematics doctorates annually in the prewar period. After 1871, new positions came into being at the reformed universities, and the École normale supérieure created a few junior positions. Beginning already in the period before 1870, under the direction of Louis Pasteur, the École normale had begun to replace the École polytechnique as the principal training ground for France's new generation of research mathematicians.

Their principal mentor during this period was Charles Hermite, who consciously emulated the German model and whose protégés included Henri Poincaré (a *polytechnicien*), Émile Picard, Paul Appell, Jules Tannery, and Émile Borel. All began teaching careers after taking a doctorate, sometimes at provincial universities, but eventually returning to Paris, where they rose to senior positions at the École normale or the Sorbonne. This system formed the basis for the rich mathematical tradition that flourished in the years before the Great War and continued beyond it. At the same time, it engendered strong loyalty to the Third Republic while intensifying the traditional rivalry between France and Germany, an attitude by no means foreign to other nations during this period of growing nationalism. Indeed, France sponsored talented students to undertake research in Germany in these decades, with the end in mind of being able to keep up with German developments. Jules Molk, one such visitor to Germany, eventually became a kind of go-between for parts of the German and French communities. Felix Klein then played a key role in internationalization of German mathematics, and Molk's most notable activity of this kind came with his partially realized translation and revision of Klein's mathematical encyclopedia.

During this period, aspiring American mathematicians necessarily turned to Europeans for inspiration (Parshall and Rowe 1994). In fact, the two most influential figures during this formative period were foreigners: J. J. Sylvester (Parshall 2006), who taught at Johns Hopkins, and Felix Klein (Tobies 2021), Professor of Geometry in Leipzig until 1886 when he joined the faculty in Göttingen. Within two years at Hopkins, Sylvester had launched the *American Journal of Mathematics*, which he edited with the assistance of William Story. At the same time, he attracted a small cadre of graduate students and associates who together comprised the first school of research-level mathematics on American soil. Klein founded no school of American mathematics, and yet his influence on the emergent mathematical research community in the United

FIGURE 2.5. The Founding Meeting of the German Mathematical Society. Photo by J. Ortgies Jr., Bremen 1890.

States was far more pervasive than Sylvester's. No fewer than six of his pupils went on to become presidents of the American Mathematical Society, and thirteen served as vice presidents. David Hilbert, too, was an important influence on mathematics in the US (and elsewhere), with influential doctoral students, including Earle R. Hedrick and Max Mason.

As various modernizing forces began to transform mathematical practice during the period from 1900 to the outbreak of the First World War, Göttingen University emerged as a model for what was to come (Rowe 2018). Under the leadership of Klein and Hilbert, this once quiet institution developed into a new kind of center for the mathematical sciences while significantly altering established norms for the conduct of research. Although a multifaceted enterprise, the Göttingen experiment was particularly influential due to the atmosphere that surrounded Hilbert's dynamic research group, which burst the mold of the traditional mathematical school (Rowe 2003). Hilbert began his career as an algebraist with strong interests in algebraic number theory, but his research interests later underwent a dramatic shift. After 1900, he and his army of students began concentrating on various topics in analysis (integral equations, calculus of variations, etc.), work strongly linked with Hilbert's interests in mathematical physics (Corry 2004a). At the same time, Hilbert's work on axiomatics was to have an important role in geometry, but also in the emerging fields of set theory, logic, and the foundations of mathematics exemplified by Bertrand Russell's and Alfred North Whitehead's *Principia Mathematica*, which appeared beginning in 1910. These trends constitute a kind of high-water mark for the perception of mathematics as *the* fundamental science, reaching out from roots in logic to cover the study of the natural world.

By 1900, the major European powers (France, Germany, Britain, Italy, and Austria-Hungary), as well as the United States, all had a professional corps of mathematicians mainly employed by their nations' institutions of higher education. Of course, these modern states were at the same time rulers of much of the world. Britain and France had large international empires, the US and Russia had expanded to cover their continents from east to west, and both Germany and Italy began to annex international territories soon after unification. These developments created new conditions that placed a premium on cultivating those mathematical sciences most needed for both military and peacetime affairs. Already in the eighteenth century, the Ottoman Empire introduced military mathematics based on European models and by the 1860s had founded its first universities. By the turn of the century, mathematics figured as a principal subject of study at Istanbul University.

A doctorate in mathematics typically provided entry into such positions, but also into teaching positions in the higher ranks of the secondary school system, as well as schools for training military officers or engineers. Those mathematicians with the requisite knowledge could also find employment as geodesists, for example at the French Bureau of Longitudes, or at an astronomical observatory, where the Gaussian tradition of predicting planetary paths had become a fine mathematical art. In an era of fast-paced industrialization and international competition, modern states became increasingly dependent on a mathematically trained workforce in many spheres of technological research and development. It was within these more applied settings that women came to play an important role in this more extended mathematical community, working as human computers long before they had the opportunity to have careers at institutions of higher learning.

Russian students seeking higher education in Western Europe came in droves, especially to Germany, France, and Switzerland. Many of them were young women, Sofia Kovalevskaya being the most famous example. Unlike Germany, Swiss universities placed no restrictions based on gender. As a result, by the early 1870s, sizable Russian colonies of female students sprang up in Zurich and Bern. Kovalevskaya spent four years in Berlin taking private lessons with Weierstrass, since most German universities were not open to women. Recognizing her unusual talent, Weierstrass arranged for Kovalevskaya to obtain a doctorate *in absentia* from Göttingen. She gained this degree in 1874 after submitting three independent research papers (Cooke 1984).

Doctoral degrees have a long prehistory, but the modern PhD first arose as part of an educational reform launched at the German universities. Over the course of the nineteenth century, faculties increasingly awarded this degree not to those who had merely mastered scholarship in a given field. Instead, they reserved this highest academic degree for those who submitted work that constituted an original scholarly contribution to their chosen discipline. In some instances—Gauss and Kovalevskaya were two such cases—dissertation research was undertaken elsewhere. The norm in mathematics, however, called for intensive study with a mentor, who often determined the student's topic of investigation. As this trend deepened, the professor's role as a teacher shifted, too. Alongside the traditional task of imparting knowledge came another, namely to train future researchers. Those who were especially adept at doing so could exert a strong influence on the direction of research they chose to practice and promote.

One of the most influential was the Berlin analyst, Karl Weierstrass, who became practically a cult figure in the 1870s and 80s. Weierstrass published little, using his lectures as the primary vehicle for presenting his latest work. Notes from his courses circulated widely, but to understand them was very difficult without attending. Adding to his mystique was the special relationship he cultivated with the brilliant young Kovalevskaya, a figure unique in the annals of mathematics. After learning that she had entered into a fictive marriage in order to study in Germany, Weierstrass realized her vulnerability and took the steps needed to secure her doctorate from Göttingen. Ten years later,

FIGURE 2.6. The Russian mathematician Sofia Kovalevskaya (1850–91). Photo by Michael Nicholson / Corbis / Getty Images.

following a tumultuous period in Russia that led to her husband's suicide, she rebounded in Sweden, where she obtained a professorship at Stockholm University. Her colleague, Gösta Mittag-Leffler, also made her an associate editor of *Acta Mathematica*, a leading international journal. In 1888, the French Academy of Sciences awarded Kovalevskaya the Bordin Prize for her work on gyroscopic motion. Then, three years later, it all ended tragically with her death at age 41. With it came an outpouring of assorted reflections about what many regarded as having been an unhappy life, a cautionary tale for any young woman who might wish to pursue a career in mathematics. Not until new research findings emerged in the 1970s and 1980s did this fundamentally negative image of her life change, leading to a renaissance of interest in this fascinating figure (Koblitz 1983).

By the decade of the 1890s, the debates surrounding Kovalevskaya's controversial career fed into a wide-ranging discourse about the role of women in modern societies. As new opportunities in higher education opened for men, a small number of women began to study mathematics at some of the elite colleges. An important pioneering figure was Charlotte Angas Scott, who studied at Girton College, Cambridge from 1876–80, after which she finished eighth in the Tripos examination, though her achievement went unacknowledged. In 1885, she obtained a doctorate from the University of London based on an external examination. That same year, Scott joined the founding faculty at Bryn Mawr College, where she taught until her retirement in 1924. Two of her seven doctoral students, Isabel Maddison and Marguerite Lehr, afterward joined the faculty at Bryn Mawr.

Maddison had studied earlier at Girton alongside Grace Chisholm. In the mid-1890s, both went on to do graduate work under Klein in Göttingen, the first German university to spearhead opportunities for (foreign) women to earn doctoral degrees (Tobies 2022). At this time, resistance to opening the Prussian universities to women was strong and widespread. Klein recalled how in 1891 the Curator of Göttingen University refused his request to allow Christine Ladd-Franklin to attend his lectures (she had studied at Johns Hopkins under J.J. Sylvester and Charles Sanders Peirce). "That's worse than social democracy," he was told, "since it merely wishes to abolish differences in property. You want to eradicate the differences between the sexes" (Tobies 2021: 412). Klein knew he had to approach this question cautiously, even though pressure was mounting to allow women into leading universities throughout Europe and the US. By this time, females had already gained the right to study at Paris University, though none had yet taken a higher degree in mathematics. Chisholm took her PhD *magna cum laude* in 1895 after passing the usual qualifying examinations. Kovalevskaya's case twenty years earlier was altogether different; she never set foot in Göttingen and earned her doctorate *in absentia*, a quite common practice that the Prussian Ministry prohibited soon thereafter.

Grace Chisholm arrived in Göttingen in the fall of 1893, along with two young American women, Mary Winston and Margaret Maltby; the latter would take her PhD in physics under Walther Nernst. Klein invited all three to attend his lectures and personally facilitated their applications for permission to attend his courses in Göttingen. This time the Ministry overruled the Curator, who resigned not long afterward. Still, the situation was anything but simple. In 1895, toward the end of her stay in Göttingen, Isabel Maddison wrote to Bryn Mawr about Chisholm's triumph, but also giving a vivid account of the kind of difficulties she and the other female auditors encountered. She reported that in the academic year 1894–5 fifteen women were attending courses in the philosophical faculty—three from Great Britain, eleven Americans, and one German—and seven of them were studying mathematics, astronomy, and physics. None, however, had been allowed to matriculate as regular students, so they were keenly aware of their status as "guinea pigs" in an educational experiment. Indeed, Madisson emphasized that:

> one should not conclude from this rapid advance in numbers that Göttingen is opening wide its doors and welcoming all comers. This is far from being the case. By no means are all the professors in favor of the admittance of women; some utterly refuse to lecture to women, others take up a neutral position, and those who are most willing to extend a University training to women are most anxious that it shall be done with discrimination, that the women students shall come well-grounded and prepared for hard work and that no element of dilettantism shall appear.[1]

Clearly, much depended on the attitudes of individual professors. Moreover, even the more liberal-minded among them took a wait-and-see attitude. This applied, in particular, to the question of admission to final oral exams. This had never been an issue with regard to German women, since they could not attend the secondary schools that administered the *Abitur*, the certificate required for admission to a German university. Once foreign women broke that barrier, the Göttingen Philosophical Faculty decided to confront this issue, though not by adopting a general policy. Instead, the professors required candidates to present their credentials individually; the faculty then granted the right to matriculate on a case-by-case basis. All three women who entered in 1893 got over that hurdle, and after passing their exams, they left Göttingen with doctoral degrees.

In the eyes of Felix Klein, at least, this experiment proved to be successful. Reporting in 1897 on six female students studying under him in Göttingen, he found their performances in every respect comparable with those of their male counterparts. These remarks appeared in a survey from 1897, in which German professors expressed their views regarding the suitability of women for university studies.[2] All four mathematicians took a positive stance toward this question, but even the more liberal-minded had obvious reservations. One respondent answered by documenting various prominent women mathematicians from antiquity to the present, but then ended with a lengthy account of Kovalevskaya's final years filled with sadness. Much of the popular literature devoted to the brilliant Russian in the years following her death had this same ambivalent quality, as myths and controversies about her life spread. For many, she stood as a warning to those women who imagined they could find happiness while competing in an intellectual field that was quintessentially suited for a select group of men. Women might entertain the idea of studying mathematics, but the notion that a woman might go on to become a research mathematician was, for nearly all at this time, almost unthinkable. Many, on the other hand, felt it was necessary to raise the standards of mathematics education for young women at the secondary schools.

Compared with Great Britain and the United States, Germany lagged far behind in the movement to create educational opportunities for young women. After 1900, various German states allowed females to take the final examination required for graduation from a *Gymnasium*, which remained an exclusively male domain. Since the curricula at the corresponding schools for females fell far below the standards at Gymnasia, candidates had to undertake intense private studies to prepare for these exams. Emmy Noether did so by studying alongside her brother, Fritz, as he prepared for his final exams, and in 1903, she passed the Bavarian *Reifeprüfung*. Four years later, she took her doctorate in Erlangen, where her father, Max Noether, taught (Rowe and Koreuber 2020).

Although a small number of post-doctoral positions were then available, these were reserved for those who hoped to move up the career ladder. Its lowest rung was an appointment as unpaid lecturer (*Privatdozent*), a title conferred after a candidate submitted a second thesis, part of a formal academic procedure (*Habilitation*) that was only open to men. Emmy Noether continued to study mathematics in Erlangen, while assisting her father and his colleagues. Her chance to pursue an independent academic career only came in 1919 (see Chapter 2 of Volume 6), the very time when women around the world were gaining the right to vote. Noether then joined the Göttingen faculty as a *Privatdozent*. Her post-doctoral thesis answered an open question about the status of "energy laws" in Einstein's general theory of relativity (see Chapter 5 of Volume 6).

By the end of the nineteenth century, students were taking lecture courses on higher mathematics at universities in Russia, Ukraine, and the Baltic states. The most prominent of these centers by far was St. Petersburg with its mathematical society founded in 1890. Four members of this group did pioneering work in number theory, probability, and applied mathematics: M. V. Ostrogradskii, P. L. Chebyshev, A. M. Lyapunov, and A. A. Markov. Kazan University was the home of one of the most famous mathematicians of the nineteenth century, N. I. Lobachevsky, who, along with János Bolyai, created non-Euclidean geometry. During the 1890s, Lyapunov was in Kharkov, where he played an important role in the Kharkov Mathematical Society, as did Sergei N. Bernstein beginning in 1908.

Bernstein had studied in Paris before leaving in 1902 for Göttingen, where he worked under Hilbert. Two years later, he submitted his doctoral thesis to the Sorbonne faculty, presumably because his French was far better than his German. Bernstein's work represents the first significant progress in solving Hilbert's 19th Paris problem. Nevertheless, when he returned to Russia in

1905, he had to take up renewed studies for a master's degree since Russian universities did not recognize degrees earned abroad. In 1908, he submitted another thesis in which he dealt with solutions to a wide class of elliptic differential equations, thereby partially solving Hilbert's 20th Paris problem. Before his appointment in Kharkov, he taught at a Women's Polytechnic College. In 1909, Otto Blumenthal urged Hilbert to find employment for Bernstein, if not in Germany then in Paris, but all such efforts proved futile. Another talented Russian Jew, Isaai Schur, managed to gain a place in the German academic world, eventually becoming a full professor in Berlin. Schur grew up, however, in a German-speaking home in Latvia. An earlier graduate of Kharkov University, V. A. Steklov, became a specialist in potential theory and partial differential equations. The Mathematical Institute of the Academy of Sciences (MIAN) in Moscow bears his name to this day. Moscow began its rise as the center of a world-renowned mathematical school only after 1900 when D. F. Egorov and N. N. Luzin began their work on descriptive set theory, measure theory, and the theory of functions (see Chapter 3 in Volume 6).

The most important of the new entrants in this internationalized mathematical world were China and Japan. In the case of Japan, following the 1868 declaration of the Meiji emperor that called for adherence to Western educational customs, Japanese scholars began to go to Europe to acquire training (Sasaki 2002). Germany was their overwhelming first choice, since the country's universities enjoyed high prestige and they were relatively easy for foreigners to attend, as opposed to the rather bureaucratic French model. Moreover, the Japanese greatly admired German military and economic prowess: already in the 1880s, their leaders promoted Prussian-style secondary schools, with militaristic rules, uniforms, and cadet training throughout Japan. In the 1890s, Rikitar Fujisawa founded a mathematical seminar based on the German model at the University of Tokyo. Fujisawa had spent five years studying in Europe, and he brought back with him the zeal to train young researchers. His most talented pupil, Teiji Takagi, studied number theory in Göttingen and later went on to contribute decisively to class field theory. The recognition of his achievements by leading Europeans marks the arrival of Japan as a major mathematical research power (Yandell 2002).

The situation for mathematics in China around 1900 was more chaotic, not least because of the social disruption that accompanied foreign incursions during the mid-nineteenth century, which ultimately led to the fall of the empire shortly before the Great War (Dauben 2002). The Boxer Rebellion of 1899–1901, aimed at foreign nationals, was one manifestation of this; remarkably, it had important consequences for the study of higher mathematics in China. This came about because the Western powers that quelled the uprising demanded reparations from the imperial government. The United States then decided to use some of these reparations to support study by Chinese nationals at American universities. A fair number of these students took first degrees and then doctorates at leading US institutions, such as Harvard and Cornell. After returning to China, some helped to establish new university programs, notably in Beijing and at Zhejiang. Thus, the Boxer Indemnity Program enabled the Chinese to replicate, in some sense, the pattern of institution building that the Americans had gained from their experiences in Germany one generation before.

The general autonomy of mathematics as a discipline, the diversification of its subject matter, the growth in numbers of its practitioners, and the maturing research environment all helped to foster the spread of national and local mathematical societies. By around 1890, these societies were holding regular meetings in France, Germany, and the United States; several were publishing their own research journals as well. The *Bulletin de la Société mathématique de France* began to appear in 1873, one year after the founding of the SMF; the *German Mathematical Society* (DMV), founded in 1890, began publishing its *Jahresbericht* that same year. Both organizations and publications continue to thrive today. Although the *Italian Mathematical Union* only came about in 1922, decades earlier Giovanni Guccia created the *Circolo Matematico di Palermo* in 1884 with its journal, the *Rendiconti.* Starting in 1888, the *Mathematical Circle of Palermo* began accepting foreign members, and by 1914, it grew into a large international mathematical society.

In other countries, mathematicians met locally, as, for example, the group that founded the London Mathematical Society (LMS) in 1865. Some twenty years later, Thomas S. Fiske, a graduate student at Columbia College, attended meetings of the LMS in the company of James Glaisher. After graduating from Columbia in 1888, Fiske and two friends decided to form a similar organization, the New York Mathematical Society. In 1891, the NYMS began publishing its *Bulletin*, and what began as a local club was renamed the American Mathematical Society in 1894. By 1902, the AMS had a vibrant Chicago Section, and soon a new San Francisco Section opened, thereby stretching its influence from coast to coast. Columbia remained the hub of its activities on the east coast after Fiske's retirement as AMS Secretary in 1896. His colleague at Columbia, Frank Nelson Cole, who edited the *Bulletin*, succeeded him as secretary. Cole had studied under Klein in Leipzig, and when he returned to the US, he became a strong advocate of research in group theory.

An even stronger impulse for new fields of research came, however, from the Chicago Section, especially the faculty at the newly founded University of Chicago, headed by E. H. Moore. Backed by the funding of John D. Rockefeller, UC boasted one of the best faculties in the country. Moore took advantage of the dearth of employment possibilities in Germany to hire Oskar Bolza and Heinrich Maschke. Both were protégés of Klein, but left their country in hopes of finding better opportunities in the New World. Moore, his two German colleagues, and Henry S. White, another former student of Klein in Göttingen, hosted a small-scale international congress, held in conjunction with a large-scale event: the 1893 World's Columbian Exposition. Its proceedings, a large part of which consisted of papers Klein solicited from mathematicians in Germany, appeared two years later, published with financial support from the AMS. In Chicago, Klein met Mary Winston and promised he would support her wish to study under him in Göttingen. Winston went on to take her doctorate there in 1896, one year after Klein's first female graduate, Grace Chisholm, completed her degree. Immediately after the Chicago Congress, a group of mathematicians convened at Northwestern University, just north of the city, where White had arranged for Klein to deliver a series of talks, published soon thereafter as the *Evanston Colloquium Lectures*. Three years later, on White's initiative, Harvard's Maxime Bôcher and Yale's James Pierpont delivered the first AMS Colloquium lectures, an honor conferred on many distinguished mathematicians since that time.

NATIONAL AND INTERNATIONAL PROJECTS, 1870–1914

As mathematicians began to define themselves through their participation in local and national societies, they increasingly chose to publicize their work in specialized journals. Leading researchers continued to submit preliminary findings or shorter works to the established European academies, but they nearly always published their lengthier studies in mathematical journals, several of which had assumed international importance by the end of the century. A noteworthy example comes from the early career of Henri Poincaré (Gray 2013), who bombarded the Paris Academy with short notes for its *Comptes Rendus,* in which he described a series of results connected with a new class of complex functions. As luck would have it, Poincaré's notes appeared just as the Swedish analyst Gösta Mittag-Leffler was about to launch a new journal, *Acta Mathematica*, with a consciously international ambit.[3] This circumstance provided the ideal opportunity for Poincaré to unveil the new theory of automorphic functions in five lengthy papers, which appeared between 1882 and 1885 in *Acta*.

By this time, visibility in mathematics largely meant presenting new research findings in major mathematical journals. Still, it was not until the mid-twentieth century that formal procedures for quality control of publications became commonplace. Before then, requests for referees' reports, whether from an editor or a member of an editorial board, were the exception rather than the rule. Since statistics on the acceptance/rejection rate during the nineteenth century are unavailable, it is difficult to gauge standards of selectivity, but even some of the more prestigious mathematical journals often accepted papers that their editors judged as mediocre. What we can judge quantitatively, however, is the approximate number of research publications that appeared in a given year, starting in 1868. This information is available thanks to the abstracting journal *Jahrbuch über die Fortschritte der Mathematik*, which counted roughly 900 published works in that year. That number grew steadily up until 1914, a trend reflected in the general growth of scientific knowledge over this period.

A far less systematic attempt to track recent literature began in 1870 in France, when Gaston Darboux founded the *Bulletin des Sciences Mathématiques*, often called "Darboux's Journal," though in some ways it resembled a newsletter. In the spring of that year, Darboux met two foreigners who happened to be visiting in Paris, Klein and Sophus Lie. Both were highly impressed by Darboux's plans for this publication, and they promised to help support it. The *Bulletin* was a monthly of 32 pages, containing four parts, two of which were dedicated to reviews of books and journal articles. It also had a bibliographical section and another with current articles or surveys. Darboux worked closely with the Bordeaux mathematician Jules Hoüel, who translated many works written in foreign languages (German, English, and Russian). Although supported by six other collaborators, Darboux and Hoüel did the lion's share of the writing and translating, a publishing venture undertaken in a truly international spirit.

Darboux was also involved in a much larger publishing venture, namely the International Catalogue of Scientific Literature, a project initiated by the London Royal Society, in cooperation with nearly all the major scientific academies on the Continent (Schappacher 2022: 38–56). Klein and Darboux attended the second London conference in October 1898 as official delegates of their respective countries. These early conferences worked out a plan for classifying the scientific literature in each of 17 different fields. The commission charged Poincaré and Klein with preparing the classification rubric for mathematics, an assignment they completed in 1901. One year later, the Royal Society finally launched this massive project, which went on for some twenty years afterward, despite the Great War, which effectively ended this and all similar cooperative ventures.

One of the most impressive international undertakings of the period was the German *Encyklopädie* project (*Encyklopädie der mathematischen Wissenschaften mit Einschluß ihrer Anwendungen*), which began in the mid-1890s but only reached completion in 1935. Published by Teubner and supported by four academies—Göttingen, Leipzig, Munich, and Vienna—the *Encyklopädie* was largely the brainchild of Felix Klein. To ensure its success, Klein exploited an international network of contacts to enlist the support of recognized authorities in all branches of mathematics and the exact sciences. By making the achievements of the past accessible through a giant compendium, he hoped the *Encyklopädie* would both stabilize mathematical knowledge and redirect the course of future mathematical research. Rather than merely acting as a compilation of facts, he saw this project as a unique opportunity to revitalize contemporary mathematics, in large part by forging closer ties with neighboring sciences.

The *Encyklopädie* was comprised of six multi-volume Bände that dealt with (1) arithmetic and algebra, (2) analysis, (3) geometry, (4) mechanics, (5) physics, and (6) geodesy, geophysics, and astronomy. This six-part structure emphasized topics at the interface between mathematics and the exact sciences while downplaying major areas in pure mathematics. Klein wanted to ensure that the traditional ties between mathematics and the physical sciences would continue to be cultivated in the future. Algebra and number theory, on the other hand, received remarkably little space, as Band 1 was not only the smallest of the six—Band 2 was three times its size and Band 3 larger still—but also by far the weakest. About half of its articles dealt with various topics from fields like probability and statistics, numerical analysis, and actuarial mathematics. As for number theory, Band 1 contains a lengthy survey article on algebraic number fields by Hilbert, but this was only a condensed version of his *Zahlbericht* from five years earlier.

Editorial responsibility for Band 4 (mechanics) fell to Klein and his assistant, Conrad Müller, and many of its articles stressed applications more than theory. Some dealt with engineering problems: ship construction, the strength of materials in machine construction (Theodor von Kármán), dynamical problems in machine theory (Richard von Mises), or the theory of hydraulic pumps and motors. Others, like the articles by Aurel Voss on the principles of rational mechanics or by Max Abraham on the geometrical foundations of the mechanics of deformable bodies, addressed theoretical issues. Paul Ehrenfest and his wife Tatyana wrote an important survey on the conceptual foundations of statistical mechanics, in which they elaborated on Ludwig Boltzmann's essay in Band 5 on the kinetic theory of gases (see Chapter 5 in this volume).

The most impressive of the six Bände was the fifth on physics (Jungnickel and McCormmach 2016), edited by Klein's former assistant, Arnold Sommerfeld (Eckert 2013). Together Klein and Sommerfeld consulted with prospective authors from foreign countries, and succeeded in attracting a number of illustrious contributors, including H. A. Lorentz and Heike Kammerlingh-Onnes from the Netherlands, and Boltzmann from Austria. Max von Laue and Wilhelm Wien joined an impressive list of authors closely associated with the Gottingen scientific community: Carl Runge, Ludwig Prandtl, Peter Debye, Max Born, Hermann Minkowski, and Arthur Schoenflies. The articles ranged over thermodynamics, molecular physics, electricity, and optics, including two classic contributions by Lorentz on Maxwell's electromagnetic theory and electron theory. Especially significant, though, was the survey article on relativity theory, written by twenty-one-year-old Wolfgang Pauli. The article on integral equations, written by Ernst Hellinger and Otto Toeplitz, only came out in volume 2 in 1927. Its delay, however, proved fortuitous, as the results it contained were highly significant for quantum mechanics. Thus, in addition to the sheer quantity of effort expended on topics involving applications of mathematics, the contributions in this area turned out to be qualitatively superior to those on pure mathematics as well.

Klein was particularly eager to win over leading foreign authorities representing three particular areas of research: British mechanics and astronomy, Italian geometry, and French function theory. He and Sommerfeld traveled to Great Britain to woe contributors, and they ultimately won over seven leading experts. A. E. H. Love wrote on hydrodynamics, Horace Lamb on acoustics, G. H. Bryan on thermodynamics, E. W. Hobson on heat conduction, G. H. Darwin on the hydrosphere, E. W. Brown on lunar theory, and E. T. Whittaker wrote about orbits and perturbation theory in dynamical systems. Equally impressive were the Italians who covered a wide variety of topics in geometry: Federigo Enriques, Gino Fano. Gino Loria, Guido Castelnuovo, Luigi Berzolari, and Corrado Segre. Altogether, they contributed nine articles between them.

Compared with Italy and Great Britain, the number of articles from French authors was comparatively meager. Georges Brunel wrote on definite integrals, and Paul Painlevé and Ernest Vessiot collaborated on a two-part article dealing with ordinary differential equations. Not coincidentally, all three of these French analysts had studied in Germany, as had the Alsatian mathematician Jules Molk, who took his doctorate in Berlin under Leopold Kronecker in 1884. Molk is remembered today as editor-in-chief of the corresponding French project, published over the period 1904–16 under the title *Encyclopédie des sciences mathématiques pures et appliquées* (Schappacher 2022: 52–4). This remarkable joint venture emerged under the auspices of the same four academies (Göttingen, Leipzig, Munich, and Vienna) and in cooperation with the publishers Gauthier-Villars in Paris and Teubner in Leipzig. Even so, the French *Encyclopédie* was hardly just a remake of corresponding parts of its German counterpart. In many instances, its articles underwent revision prior to translation; in others, an entirely new survey was prepared for the French edition. Thus, while its overall structure still followed that of the German model, the contents of the *Encyclopédie* volumes often differed markedly from the original versions.

Molk oversaw the first three volumes, supported by a number of distinguished French mathematicians. For the section on arithmetic in volume 1, he gained additional contributions from Jules Tannery, Élie Cartan, Maurice Fréchet, and René Baire. A new section in volume 2, edited by Émile Borel, presented recent research on the theory of real functions, a topic barely touched on in the German *Encyklopädie*. Another major improvement appeared in volume 3 with Cartan's expanded version of the article by Gino Fano on Lie groups.

Other important contributors to these three volumes included Paul Appell, Jacques Hadamard, and Édouard Goursat. Cooperation between the French authors and their German colleagues apparently went smoothly so long as it lasted. Molk also enlisted the support of the Swedish historian of mathematics, Gustaf Eneström, who collaborated with Paul Tannery in strengthening the quality of the historical information in the French edition, which was more extensive and reliable than in the original German edition. The fourth volume, edited by Paul Appell, also contained new supplementary material: Borel on statistical mechanics, Gabriel Koenigs on kinematics, Paul Langevin on deformable systems, and Roger Liouville on recent French ballistics studies. Liouville's article was among the last to appear in the *Encyclopédie*, a fitting topic no doubt for putting an end to Franco-German cooperation on the brink of the Great War. Molk died in Nancy only a few months before the guns of August 1914 sounded. Parts of the volumes on physics, geodesy, and astronomy came out during the first two years of the war, but afterward the project came to a premature close.

Klein's original aim had been to reintegrate pure and applied mathematics, a goal that with the advantage of hindsight was hopelessly ambitious. Already around 1900, mathematics as a whole was in the throes of a profound structural transformation and reorientation that would take decades to unfold. We described this tectonic shift in the introduction to this volume as one in which classical mathematics gave way to modern mathematics. Many of the earlier articles in the original *Encyklopädie*, which went on until 1935, were long obsolete by the time this massive project ended. In view of this circumstance, a new effort began during the NS-era, and over the years from 1939 to 1959 Teubner published another 13 supplementary volumes. Nearly all of them dealt with topics in pure mathematics, including modern algebra, number theory, and logical foundations.

In the meantime, the first Soviet Encyclopedia of Mathematics had appeared with a second version underway, eventually completed in the 1950s. Other similar projects arose in France (see Chapter 3 of Volume 6) as well as Japan, and the vibrant US American mathematical community later translated works from these two countries into English. All these compendia struggled to keep pace with the most recent advances, whereas the German mathematicians who tried to keep the *Encyklopädie* alive lacked the resources to compete with these other projects. In the aftermath of the Nazi era, mathematics made in Germany lost its former importance and overall allure.

The two decades that preceded the Great War also witnessed the most visible form of international cooperation among mathematicians, namely the staging of international congresses (Lehto 1998). The first International Congress of Mathematicians (ICM) took place in Zurich in 1897, with just over 200 attendees. Switzerland, homeland of Euler and the Bernoullis, also enjoyed a special status in European affairs as a multicultural country, and this made Zurich an ideal venue for this groundbreaking event. At the opening banquet, Professor Adolf Hurwitz at the Swiss Federal Institute of Technology (ETH), remarked:

> It is true that the great conceptions of our science generally develop and mature in the silent study of the scholar. No science, unless perhaps philosophy, has such a brooding and solitary character as mathematics. Yet nevertheless, there lives in the breast of the mathematician the necessity for communication, for conversation with colleagues. (Albers et al. 1987: 4)

Another principal organizer of the Congress was Hurwitz's colleague, the historian of mathematics Ferdinand Rudio, who served as General Secretary and editor of the proceedings. In his opening address, Rudio took the opportunity to inform his audience about plans to publish the *Complete Works* (*Opera Omnia*) of Leonhard Euler. Various efforts to launch this project had occurred throughout the nineteenth century, including a planned joint venture envisioned by the Academies of St. Petersburg and Berlin, the two institutions where Euler had spent nearly his entire career. In the end, however, the two sides could not reach agreement about how to proceed. A decade after the Zurich Congress, Rudio achieved an important breakthrough in Basel at the celebration of the bicentennial of Euler's birth. This led to the founding of the Euler Commission, an organization that still exists today. This Commission launched a highly successful subscription fundraising drive that raised close to 500,000 Swiss Francs. In 1911, the first volume in Euler's *Opera Omnia* rolled off the presses, his classic Vollständige *Anleitung* zur *Algebra*, edited by Heinrich Weber. By 1914, twelve volumes in this giant edition were already in print.

The general success of the Zurich ICM sparked an ongoing series of such meetings that has continued, interrupted only by the World Wars, up to the 2018 ICM in Rio de Janeiro, the first in a South American country. The 2022 ICM had been scheduled to take place in St. Petersburg, but had to be held as a virtual event after the Russian invasion of Ukraine. Much like the modern Olympic Games—which began in 1894, when Baron Pierre de Coubertin founded the International Olympic Committee (IOC) that led to the first modern Games in Athens two years later—the ICMs occur every four years. This quadrennial pattern began with the second ICM, which took place in Paris in 1900 amid the chaos of the city's International Exposition in that year. The event itself was hardly memorable except for an address on "Mathematical Problems" that Hilbert delivered to an attentive audience at the Sorbonne (Gray 2000; Yandell 2002). In it, he unveiled ten of the famous Hilbert Problems; the remaining thirteen soon followed in published form. Fittingly enough, Klein's former student, Mary Winston Newson, made Hilbert's text

FIGURE 2.7. Advertisement for the First International Congress of Mathematicians held in Zürich in August 1897. Image courtesy of Guillermo Curbera.

with the complete list of 23 problems available to the still fledgling American mathematical community in 1902, when she published an English translation in the *Bulletin of the New York Mathematical Society*. The success of this address owed much to the venue and year, but also Hilbert's picture of mathematics as a larger enterprise consisting of interrelated disciplines. His problems were anything but isolated curiosities. He saw them instead as pivotal questions tied to various subfields, a vision closely linked with a structural understanding of mathematics that was beginning to shape the outlook of several future leaders of the community.

The prestige of the international congresses only grew stronger in the years leading up to 1914. The localities—Heidelberg (1904), Rome (1908), and Cambridge (1912)—adequately reflect the mathematical prestige of these host nations, whereas participation from around the world gradually expanded. If the mathematics undertaken in this international setting remained tied to various national and local programs of research, the larger community was nonetheless well established, sharpened by keen national rivalries. The congresses tended overwhelmingly to showcase results in pure mathematics, though applications in physics and other fields were by means rare. Mathematicians working outside the university framework, if they participated in these professional communities, usually did so in the context of the specialized fields in which they were active. These patterns remained highly stable up until the outbreak of the Great War, when quite suddenly everything changed. It would take several decades to restore harmonious relations between mathematicians in the leading national communities.

CHAPTER THREE

Inventing Mathematics

Rigor, Creation, and the Unexplored

JEREMY GRAY

The profound changes that mathematics underwent in the long nineteenth century can usefully be reviewed under three headings: new subjects, better proofs, and different answers. Included among the new subjects are topics which became mathematized that were not simple extensions of what was already known. The idea of better proofs refers to a new, more abstract and autonomous sense of rigor that changed what could be said and what had to be done. Improved rigor was not an insistence on spelling out the obvious, but a tool for making new discoveries and for moving mathematics away from a reliance on naive intuition—one reason higher standards of rigor were required was to legitimize some of these new research areas. Different answers refers to answers of a qualitatively different kind, often more conceptual and less explicit, answers that might, for example, prove that a certain equation has a solution but do nothing to exhibit it explicitly (or show that an equation has no solutions of a given kind at all). It will become clear that the ideas of better proofs and different answers are linked, much in the way that new means of discovering things lead to new kinds of things being discovered.

The net effect was to make mathematics a more self-contained, autonomous, and abstract subject, one that was increasingly independent of contemporary science and had its own criteria for merit, its own definitions, and its own methods. Loosely, one could say that it made mathematics more "pure," or perhaps "pure" in a different way, but ultimately it is not helpful to try to divide the subject into "pure" and "applied" halves. The reasons for these changes go beyond the inclinations of a few influential mathematicians and into the creation of a mathematics profession and the evolving study of physics, and beyond that into broader social changes. These are discussed elsewhere in this volume: here we concentrate on the invention of new mathematics in the long nineteenth century.

To establish the importance of new standards of rigor we will look chiefly at how the calculus became rigorized. Mathematical analysis, as the rigorized calculus became known, can be said to start with Augustin Louis Cauchy, who reformulated the ideas of his eighteenth-century predecessors, but only to reach a stable and generally accepted form by the end of the century with the work of Karl Weierstrass, Richard Dedekind, and others. We can look at this process here only selectively. One particularly interesting area is its interaction with the ideas of Joseph Fourier, who claimed to have proposed a dramatic and highly effective way of representing *any* function as an infinite series. The extent to which he was correct and his representations actually useful was to remain unclear for a long time, as we shall see.

The clearest examples of mathematicians creating new fields come from algebra and geometry. In algebra the theory of numbers was wholly redefined by Carl Friedrich Gauss in a book of 1801. Thirty years later, the theory of polynomial equations was given wholly novel foundations by Évariste Galois, and once these foundations were accepted, in the 1860s, they became a cornerstone of the new theory of groups. Likewise, in geometry, the description of a radically different but physically plausible geometry of space that differed from Euclidean geometry, the so-called non-Euclidean or hyperbolic geometry, also took a generation to be accepted, but it helped open the way to further novel formulations of geometries of many kinds when it was connected to Gauss' reformulation of the geometry of surfaces.

By the end of the century, the drive for autonomy and rigor was leading to surprising, and sometimes contentious new branches of mathematics, often of an axiomatic kind, and often with novel foundations in an emerging theory of sets. This set the stage for the production of a distinctively modern mathematics in the twentieth century.

RIGORIZING REAL ANALYSIS

The usual story of the calculus, as we saw in Chapter 3 of Volume 4, draws attention to growing concern among Lagrange and others in the years around 1800 that the calculus lacked adequate foundations. As argued in (Grabiner 1981), and described more fully in (Schubring 2005), these attempts contained many good ideas that Cauchy was able to draw on in order to present his own much more fruitful reworking of the subject, a presentation that proved capable in the hands of later mathematicians of finally establishing the calculus soundly.

But mathematicians did not invent the calculus for its own sake. It was not a game of finding integrals and differentiating elaborate expressions—or at least, not solely. The calculus was essential in celestial mechanics, the motion of fluids, the geometry of curves and surfaces, and other topics of wider interest. The journey from Newton's *Principia* to Laplace's *Mécanique Celeste* made heavy use of the approximate solution of systems of ordinary differential equations. Success in this endeavor inspired most of the leading mathematicians of the eighteenth century to attempt problems involving more than one independent variable, and so the subject of the calculus of several variables was born, and with it the study of partial differential equations.

All eighteenth- and early-nineteenth-century accounts of the solution to a differential equation share a common feature that was both productive and a hinderance: the solution method is to be explicit. It should be possible in principle to follow it through in any given case to a solution in a form that could give precise numerical answers, although practical problems might deny you complete success. This could happen if the solution could only be given as a power series, or in the form of an integral that was impossible to evaluate.

The good thing about this approach is that it focused the attention of mathematicians on finding useful answers. The bad thing was that it obscured the question of whether the equation had a solution at all. Could it be that the failure to express a solution as a formula arose from the fact that the equation might have *no* solution?

In 1819, Cauchy published his account of the general first-order partial differential equation. In many ways it resembles Monge's account of 1808, which Cauchy could well have met when a student at the École polytechnique. But Cauchy noted that his method extends to any number of independent variables, which Monge's did not, and so it establishes the existence of solutions, under prescribed conditions, to these equations, too.

Perhaps the next year, the date is uncertain, Cauchy took a much more radical step in lectures at the École polytechnique, where he now taught, and established conditions that ensure that an ordinary differential equation, has a solution. More precisely, he showed that the equation

$$\frac{dy}{dx} = f(x,y)$$

has a solution that takes a given value y_0 at a given point x_0 if the function f satisfies certain explicit conditions in some neighborhood of the point (x_0, y_0) in the plane.

His method could be seen as little more than spelling out what is needed to make Euler's method of approximating the solution into a proof that it works—but it is much more than that. It is a *proof* that the solution exists, and by

FIGURE 3.1. Augustin-Louis Cauchy (1789–1857). Photo by Betmann / Getty Images.

implication a suggestion that a solution might not exist if these conditions are violated (and indeed that is sometimes the case).

Cauchy followed this with his famous, if nonetheless in some ways obscure, rigorization of the calculus. In his lectures of 1821 at the École polytechnique, published as his *Cours d'analyse* (Cauchy 1821/1990), Cauchy explained with unusual precision what he meant by such concepts as variable quantities (including infinitely small and infinitely large quantities), real functions and their continuity, and convergent and divergent series. He defined the crucial concept of limit this way:

> When the values successively attributed to the same variable indefinitely approach a fixed value, in such a way as to finish by differing from it by as little as one wishes, this last is called the *limit* of the others. (Bottazzini 1986: 103)

He defined an infinitely small number as a variable taking successive numerical values and decreasing indefinitely in such a way that it becomes smaller than any given number. The concept of an *infinitely small* quantity, α, of various orders α^n was important for Cauchy—it is how he expressed many arguments involving limits—and he gave rules for handling them, noting that one must distinguish between decreasing constantly and decreasing indefinitely.

A function f—defined as a variable that depends on another variable—was said to be *continuous between certain limits* if, for each x within those limits there is a unique, finite value of $f(x)$ and the absolute value of $f(x + \alpha) - f(x)$, where α is infinitely small, decreases indefinitely with α. We note that Cauchy defined continuity on an interval, not at a point, which was to prove the source of some later confusion, but he did note that continuity can fail at a point. To base the calculus on the concept of continuity was a radical step to take at a time when almost all mathematicians of the time took differentiability for granted: continuity had never been given a prominent position in the edifice of analysis before. To make it work, Cauchy took some ad hoc techniques of Lagrange, notably those for dealing with infinite series, and elevated them to foundational status. Cauchy also took more seriously than his predecessors the idea that an infinite series $\sum_{n=0}^{\infty} a_n x^n$ might converge only for values of x in a certain interval, or for no non-zero value of x at all, and he gave several tests for the convergence of infinite series.

Precise definitions make it possible to give precise proofs, and among the important results that Cauchy proved is what is called the *intermediate value theorem*, which says that if a function $f(x)$ is continuous between $x = a$ and $x = b$, and c lies between $f(a)$ and $f(b)$ then there is at least one x between a and b such that $f(x) = c$. In other words, a continuous function on an interval that is positive somewhere and negative somewhere else must be zero somewhere in between. Cauchy's proof was a great advance over previous proofs that appealed to the continuity of the graph of the curve in some loose, geometrical and wholly intuitive way.

Then, in his *Résumé* of 1823, Cauchy showed how a theory of differentiation and integration could be built up, based on his concepts of limit and continuity. He defined what it is for a function to be differentiable at a point, and what the indefinite integral of a function is. He defined the integral as a limit of a sequence of approximations to it, and not, as had become customary, by defining integration as the inverse of differentiation, so he then had to prove the Fundamental Theorem of the Calculus. This says that if $f(x)$ is a function defined on an interval and its indefinite integral $\int_a^t f(x)\,dx$ exists, then the integral is differentiable, so indeed integration and differentiation are inverse processes:

$$f(t) = \frac{d}{dt}\int_a^t f(x)\,dx.$$

Cauchy strove for higher standards of rigor than his predecessors. In the preface to his *Résumé* he quite reasonably claimed,

> [t]he methods that I have followed differ in many respects from those found in works of the same kind. My principal aim has been to reconcile rigor, which I had made a law unto myself in my *Cours d'analyse*, with the simplicity that results from the direct consideration of infinitely small quantities. (Bottazzini 1986: 118)

The organization of the *Cours* and the *Résumé* is much more coherent than previous treatises, the chain of arguments is both longer and more conspicuous, and much more is actually *proved* than was the case in earlier accounts, which took many things for granted or regarded them as simple facts. Indeed, it sometimes showed that results that had been widely assumed might in fact be wrong.

That said, his work left a lot to be done and quite a lot to be desired. He made some mistakes, he had a bad habit of introducing hypotheses into a proof that do not appear in the statement being proved, he often put things in words that would have been clearer in symbols. But the rigorous calculus is still based on the concept of limit, out of which are spun the concepts of integrability, continuity, and differentiability, and the chain of theorems that connect them. One might say that whereas previous mathematicians took it for granted that functions are continuous, can be differentiated, and have integrals, Cauchy defined these terms and henceforth required it to be shown of any function whether it was continuous, integrable, or differentiable.

As had been the case with ordinary differential equations, the solutions to partial differential equations in the eighteenth century were sought in the form of a formula involving the independent variables. But instead of the solution being adapted to the initial information for one value of a single independent variable (usually time), it must be now adapted to the initial values of all the independent variables, and so, it was thought, the solution should be a general expression or formula with some arbitrary functions in it that would be determined from these the initial conditions. The wave equation was exemplary (see Chapter 3 in Volume 4). But the Laplace equation was a worry: a few solutions were known piecemeal, it was not clear that they could represent an arbitrary solution, and there was no reason to believe that the problem could be solved for arbitrary initial conditions.

Even less was known about the equation that Fourier had discovered was the equation for the propagation of heat along a line. Doubtless inspired by the example of the wave equation, he claimed in his book *The Analytical Theory of Heat* (1822) that every solution was a possibly infinite sum of terms, each of which was of the form of a constant multiple of cos nx or sin nx multiplied by e^{-n2t}. So, the solution will be of the form:

$$\sum_{n=0} a_n \cos nx e^{-n^2 t} \text{ or } \sum_{n=0} b_n \sin nx e^{-n^2 t},$$

Claims like this had been made before. What distinguished Fourier's work was that he argued that whatever the initial conditions might be (whatever the distribution of heat is on the boundary at some initial time) the constants and a_n and b_n can be determined so that the infinite sum is a solution, and he gave a simple formula for determining them. He even offered a proof that *any* function $f(x)$ can be written in the form:

$$\sum_{n=0}^{\infty} a_n \cos nx \text{ or } \sum_{n=0}^{\infty} b_n \sin nx,$$

or as a sum of these two infinite series.

It is worth pointing out just how shocking these claims are. These series can represent functions with jumps; one of Fourier's examples was the series for the function that is $+ \pi/4$ on the interval $(-\pi/2, \pi/2)$, $-\pi/4$ on the interval $(\pi/2, 3\pi/2)$, and so on:

$$\cos(x) - \frac{1}{3}\cos(3x) + \frac{1}{5}\cos(5x) - \frac{1}{7}\cos(7x) + \cdots$$

Each term in the infinite sum is an infinitely differentiable function, so plainly the infinite sum lacks many of the properties of its components. But why then it is obvious that Fourier's rule for finding the coefficients works, which assumes that exchanging the order in which summation and integration are performed does not matter, true though that would be for finite sums? Is it the case that the infinite sum that supposedly represents a given function is actually equal to it? Indeed, is it obvious that the infinite sums Fourier required even converge and make numerical sense?

Fourier was a generation older than Cauchy, and his idea of rigor was of his time. Once Cauchy had begun to show what was required of the rigorous calculus, all Fourier's claims were called into question. First Siméon Denis Poisson, then Cauchy looked at the issue and sought to give more rigorous accounts, but the first to succeed was a young German visiting Paris: Peter Gustav Lejeune Dirichlet (Merzbach 2018).

In a small masterpiece of rigorous analysis, sharper than anything Cauchy was ever to produce, Dirichlet was able to show that if the function $f(x)$ is continuous and monotonic (either never decreasing or never increasing) on an interval then its Fourier series converges and is everywhere equal to the value of the function. In other words, the function is represented correctly by its Fourier series. The conditions he had to impose on the function are significant; that any conditions had to be imposed was unexpected. That the proof explicitly used these restrictions is typical of how the calculus was used after Cauchy:

properties of functions and the validity of arguments about them should be grounded in explicit assumptions, and answers should be qualified, giving assurance that something happens if something else does, but not necessarily otherwise.

Dirichlet's work began a series of investigations which were not to be finished for well over a century, that ask for the study of the most general class of functions to which Cauchy-style mathematical analysis allows things to be said. Growing skepticism about Fourier's sweeping claims promoted not only higher standards of rigor. It also advanced the capacity of rigorous mathematics to highlight essential differences and to discover novel functions with highly unintuitive properties (such as functions that seldom if ever agree with their Fourier series!). It is not only that Fourier's claims lacked proof, their naivety hid a great deal of substantial and surprising mathematics, including the nature of the real numbers, of sets of real numbers, and indeed set theory itself. His eighteenth century formulations eventually gave way to qualitatively different, late nineteenth century, abstract mathematics.

ALGEBRA

The theory of numbers was dear to Fermat, but he had failed to interest his contemporaries in it. As we saw in Chapter 3 of Volume 4, it was brought into the concerns of the mathematical community by Euler, Lagrange, and Legendre. But the whole subject was then rewritten by Carl Friedrich Gauss in his book the *Disquisitiones Arithmeticae* of 1801, published when he was only 24. In it, he gave the subject the form it was to keep for the nineteenth century and even beyond (see Goldstein, Schappacher, Schwermer 2007).

Certainly Gauss gave more rigorous proofs of some of the elementary results in the subject, and he was not afraid of pointing out when his illustrious predecessors had failed. But he did much else. A central question in the subject concerns what are called *binary quadratic forms*—expressions of the form $ax^2 + 2bxy + cy^2$ in which a, b, and c are integers—and the values they can take when x and y are integers. For example, there are integers such that $x^2 + y^2 = 5$ (for example $x = 2$ and $y = 1$) but there are no integers such that $x^2 + y^2 = 7$. A lovely theorem, claimed by Fermat but first proved by Euler, says that an odd prime number can be a value of $x^2 + y^2$ if and only if the prime number is of the form $4n + 1$, and moreover that an integer is a possible value if and only if any odd prime factor of it of the form $4n - 1$ occurs to an even power. There are similar theorems about other quadratic forms, such as $x^2 + 2y^2$ and $x^2 + 3y^2$, but already $x^2 + 5y^2$ causes problems, because it cannot take the values 3 or 7 but it can take the value 21 (let $x = 1$ and $y = 2$). A general theory that would say what values $ax^2 + 2bxy + cy^2$ can take for given a, b, c was entirely lacking, and that is what Gauss went a long way toward providing.

Lagrange had described how changes of variable can be used to reduce any given quadratic form to its simplest version without changing the value of its discriminant $D = b^2 - ac$. It follows from his theory that every quadratic form with a given discriminant D can be reduced to exactly one of a finite number of quadratic forms with that value of D, none of which can be reduced to any other. (For example, when $D = -5$ the only possibilities are $x^2 + 5y^2$ and $2x^2 + 2xy + 3y^2$.) He also showed that if the quadratic form takes the value m then D must be a square modulo m, i.e. there exists a number x such that $x^2 - D$ is divisible by m. In the example of $x^2 + y^2$, $D = -1$ and -1 is a square modulo 5 but not modulo 7, as you can easily check. So a crucial theorem in the subject (as we shall see later) became the question of when a prime number p is a square modulo another prime number q.

FIGURE 3.2. Carl Friedrich Gauss (1777–1855). Photo by Fine Art Images / Heritage Images / Getty Images.

The way forward was to reinterpret this question. Let us take the primes $p = 5$ and $q = 7$. The non-zero squares modulo 5 are 1 and 4, and the non-zero squares modulo 7 are 1, 4, and 2. We see that 5 is not a square mod 7, and 7 is not a square mod 5. In general an important theorem says that, for odd primes p and q, p is a square modulo q if and only if q is a square modulo p unless both primes are congruent to -1 mod 4, in which case p is a square modulo q if and only if q is not a square modulo p. It is easy to check this theorem in any given case, but proving it was much harder, and Gauss noted that the only substantial attempt at a proof (due to Legendre) was badly flawed. He then gave two proofs in his *Disquisitiones Arithmeticae*, and went on to publish four others; two more were left unpublished.

The theorem just mentioned is called the theorem of quadratic reciprocity (quadratic because it is about squares, reciprocity because the primes p and q play reciprocal roles). Gauss used it to discover and prove a remarkable fact. When the quadratic forms with a given value of D are considered (arranged in their simplest form, which is subtler than Lagrange's classification) and one asks which can take a given odd value m the answer is precisely half of them. This was a very difficult result to prove, and it changed the nature of the subject entirely. It can be argued that this is the first theorem in mathematics that other mathematicians called "deep," meaning profound, hidden, and crucial in shaping the subject.

Gauss' other proofs of quadratic reciprocity further contributed to the sense that number theory was deep, because they exposed surprising connections with other parts of mathematics, and indeed seemed to require them. Gauss himself commented that theorems in the subject are easy to spot but often very difficult to prove.

The net effect was that number theory after Gauss became the branch of mathematics that no mathematician could afford to ignore, or at least no German mathematician. Dirichlet said he was never without a copy of the *Disquisitiones Arithmeticae*, and made it his business to try to explain it to others. In so doing he discovered some other deep results, most famously that in every arithmetic progression $a, a + b, a + 2b, \ldots, a + nb, \ldots$ with a and b relatively prime there are infinitely many primes.

In Gauss' hands number theory became a new subject, much more abstract and remote from recreational mathematics. Lagrange had shown that a surprising amount of the study of quadratic forms can be done with elementary algebra. But the theorem of quadratic reciprocity, and above all the result that only half the possible quadratic forms occur, showed that elementary algebra was superficial, and new concepts were required to find the best results. Also, Gaussian number theory rested on better proofs. Not only were they rigorous, they were conceptually deeper. Finally, its questions called for new types of answer, about the structure of, as it might be, all quadratic forms of a given discriminant, and here Gauss proved the remarkable fact that quadratic forms of the same discriminant can be multiplied together and form, as we would say, a group. Mathematics has had a lot to say about groups, but in 1801 the idea that classes of objects could be multiplied together was shockingly novel.

Algebra, which in schools has a lot to do with solving polynomial equations, also changed greatly in the nineteenth century. Since Mesopotamian times every school child has had the chance to know how to solve the quadratic equation $ax^2 + bx + c = 0$ by the rule summed up in the formula for the solutions. Similar, though much more complicated, formulae were found for cubic and quartic equations (see Chapter 3 of Volume 3), but not for the general equation of degree 5. Lagrange eventually began to suspect that this was because it was impossible: there could be no formula involving the elementary operations of arithmetic and the extraction of roots. (An equation with a formula of this kind for its solutions is said to be *solvable by radicals.*)

It was to turn out that Lagrange was correct. The first to get close to a proof was the Italian mathematician Paolo Ruffini; his long argument is not entirely rigorous but it was well on the right lines. Independently and a little later, in 1826, the Norwegian mathematician Niels Henrik Abel gave a much shorter and more rigorous proof. He went on to show that at least some of the polynomial equations of degree greater than or equal to 5 that turn up in some other branches of mathematics are nonetheless solvable by radicals. Abel died young, but his work acquired one remarkably perceptive reader, the Frenchman Évariste Galois (Gray 2018: 115–42).

What is involved in solving a polynomial equation with a formula of this kind? The solution formulae, when they exist, can be complicated. For example, the formula for the solution of a cubic equation involves cube roots of an expression that involves square roots. Abel's approach started by saying that any formula of the required kind for the solution of the given equation would be produced by taking the coefficients of a given equation and combining them (by addition, subtraction, multiplication, and division) and then taking nth roots, in a process that could be repeated finitely many times (for various n). The result would be a formula that involves roots of an expression that itself involves roots, and so on. The hope is that after doing this a finite number of times one obtains a new number equal to one of the roots of the given polynomial equation.

In a little more detail, at the end of the first stage one has a rational number q_1 formed from the coefficients of the equation, and one takes its n_1th root, β_1. This gives us a collection of expressions involving the rational numbers, β_1, and its first $n_1 - 1$ powers that can be added, subtracted, multiplied, and divided; let us call it K_1. One then forms some fractional expression in these new numbers of which one takes the n_2th root. This gives a new collection of number-like objects, call it K_2, and so on.

Galois' idea, drawing on the work of Abel and possibly Lagrange, was to focus on the transformations of K_1 to itself that leave every rational number fixed. These transformations form what we call a group—the word was in fact introduced by Galois, but with a somewhat different meaning. He then focused on the transformations of K_2 to itself that leave every member of K_1 fixed, they too form a group, and so on. It turns out that all these groups are cyclic. So the process produces a sequence of cyclic groups.

On the other hand, as Galois showed, the permutations of the roots of the given equation also form a group, so in his hands the question became: Can the group of the given equation be obtained as a sequence of extensions of one cyclic group after another? His answer was, of course, No, in general. Therefore there are polynomial equations of degree 5 than are not solvable by radicals. His argument was more complicated than can be described here, in particular it required a novel property of groups, but in essence his theorem was that a polynomial equation is solvable by radicals if and only if its group can be constructed out of a sequence of cyclic groups. This was an advance on all previous work because it gave a criterion applicable to any polynomial, and so explained why some equations of degree 5 or more are solvable by radicals.

The disadvantage of Galois' method was not so much that his criterion would be hard to use as that the concept of a group was new and was poorly explained by him. In particular, he required that the group of the equation contain a subgroup of a special kind. This was a step too far for his generation of mathematicians, and Galois' ideas languished until the 1860s. It did not help that some of his key ideas were only explained in a memoir that he wrote before going to a duel in which he received fatal injuries; he died at the age of twenty.

Eventually, Galois' ideas completely transformed the subject: after the 1870s the solvability of polynomial equations became part of a theory of groups that was exceptionally fertile (see Wussing [1968] 1984; and Gray 2018). That every polynomial comes with a group means, in many areas of mathematics, that the important properties of the polynomial are read off from the properties of its group. In this sense—and even though it cannot be said that Galois himself provided better proofs—the answers provided by Galois' approach were better: more conceptual, more adaptable to other contexts in which polynomials arise, anchored in a theory of groups that was developed independently. It is for these reasons that his work is honored in an unusual way: it is called after him—Galois theory (Corry 2004b).

GEOMETRY

During the eighteenth century, widespread appreciation of the merits of Euclid's *Elements* came with a disquiet about some of its aspects, most notably the parallel postulate. Given the other assumptions made in the *Elements* the *parallel postulate* is equivalent to the claim that given a plane, a line ℓ in the plane, and a point P in the plane that does not lie on the line, there is a unique line ℓ' through P that does not meet ℓ. This line is called the parallel line to ℓ through P, and most of the results in the *Elements* depend on it.

But is it true? Or, if you prefer, is it necessary to assume it? It had been shown in the eighteenth century that one could not assume that there are no parallels and every line through P in the plane meets ℓ. This is inconsistent with Euclid's other assumptions. That left open the alternative that there is more than one line (and hence infinitely many lines) through P that do not meet ℓ. This too had been investigated, and many surprising consequences deduced from it, but no actual contradiction with the rest of the *Elements*. Notably, the sum of the angles in any triangle in this geometry must be less than 180°.

Then around 1830 János Bolyai in Romania–Hungary and Nikolai Ivanovich Lobachevsky in Kasan in Russia independently offered a strikingly similar new approach to geometry and the resolution of the parallel postulate (Gray 2011). Lobachevsky's was in some ways more fundamental, and we shall largely confine this account to his. He took the concept of distance as primitive, and using it defined spheres and planes in three-dimensional space. So, given two points in space he considered two spheres of the same radius, one centered on one point and one on the other. When these spheres meet they meet in points equidistant from the two spheres, and as the common radius varies these points fill out and so define the plane equidistant from the two points. A similar argument defines straight lines in the plane.

Both Bolyai and Lobachevsky supposed that in a plane with a line ℓ and a point P not on ℓ the lines through P that lie in the plane divide into three families:

- lines that meet ℓ
- two lines, one in each direction, that approach ℓ asymptotically
- lines that never meet ℓ and diverge from it in each direction.

The two lines asymptotic to ℓ they called the (directed) *parallels* to ℓ through P; one in each direction. Suppose now that the perpendicular from P to ℓ meets it at Q. Then the angle each parallel through P to ℓ makes with PQ depends on the length PQ; the function that specifies this dependence Bolyai and Lobachevsky called the *angle of parallelism function*. This function was crucial to the work of both men.

Both men also considered families of directed parallel lines: the family of all lines in a plane parallel to a given line in the same direction. They then considered the curves that cross a family of directed parallel lines at right angles, which Lobachevsky called *horocycles*. In Euclidean geometry, these curves are the straight lines at right angles to a family of parallel lines, but in the new setting they are different and more interesting. They were also able to obtain a formula that says that the parallel lines diverge exponentially from each other (whereas in the Euclidean case the parallel lines remain equidistant).

The same process can be followed in three dimensions. Take the family of all lines in space parallel to a given line. It is cut at right angles by a family of surfaces, which Lobachevsky called *horospheres*. Select one of these horospheres. Any two of the family of parallel lines lie in a common plane, and this plane cuts the horosphere in a line. So any three of these parallel lines (that do not lie in a common plane) define a triangle on the horosphere. What Bolyai and Lobachevsky both showed is the extraordinary fact that the angle sum of such a triangle on the horosphere is 180°—the geometry on the horosphere is Euclidean!

Then, by working in three dimensions and using results that they had been able to obtain for the angle of parallelism function, they established the formulae of hyperbolic trigonometry, that connect the sides and angles of triangles in two-dimensional non-Euclidean space. The formulae are strikingly analogous to the formulae that apply to triangles on the sphere. They also show that small triangles in the new geometry would be almost indistinguishable from Euclidean ones, which would explain why non-Euclidean geometry had never been detected and only astronomical measurements could decide whether Euclidean geometry or the new geometry correctly described the nature of physical space.

Both Bolyai and Lobachevsky said that the passage from geometrical concepts to the formulae showed that the formulae were geometrically meaningful and therefore consistent, and so the new geometry was logically coherent. This argument may not be wholly convincing—although some distinguished mathematicians later accepted it—but it was not imperfections in the work of Bolyai and Lobachevsky that explained its dismal reception. With one notable exception it was simply rejected out of hand, and gained no recognition until after both men were dead. It is hard to explain this as deriving from anything other than the relative obscurity of its two discoverers and a wholesale unwillingness to contemplate the fact that there could be a geometry other than Euclid's.

The exception was Gauss, the most distinguished geometer of the time, but he accepted the new geometry only in private and not in a way that helped either Bolyai or Lobachevsky advance their cause. Gauss was a friend of János Bolyai's father Farkas, but the letter he wrote to them when he read János' work (which had been published as an appendix in his father's two-volume textbook on geometry) made János decide never to publish again. Gauss did propose Lobachevsky for membership of the Göttingen Academy of Sciences, but never otherwise spoke up for him. There can be little doubt that had Gauss endorsed their findings at the time their careers, and the progress of geometry, would have greatly benefitted.

We know this because the discovery of unpublished papers left behind at Gauss' death is precisely what made non-Euclidean geometry acceptable. As the news spread that Gauss had agreed with Bolyai and Lobachevsky, and done work of his own on the new geometry, a generation of mathematicians took it up, improved it, and exploited it. We shall look at some of that below, but before then we should look at what Gauss' own contributions to geometry were, because they changed the subject in ways that eventually made the discoveries of Bolyai and Lobachevsky look like a special case.

Differential geometry is the name given to the study of geometry by means of the calculus, and this had been done ever since the calculus was invented. Mathematicians of the late seventeenth century knew that lengths along curves in the plane are evaluated as the integral of the element of arc length *ds* by using Pythagoras' theorem in the form:

$$ds^2 = dx^2 + dy^2.$$

The corresponding way of measuring the length of a curve on a surface in space with equation $z = f(x, y)$ uses Pythagoras' theorem in three dimensions:

$$ds^2 = dx^2 + dy^2 + dz^2.$$

The result is a simple formula involving dx, dy, and the partial derivatives of f with respect to x and y.

In the eighteenth century Euler had begun the exploration of the differential geometry of surfaces, and offered a way to define the curvature of a surface, but the modern theory of differential geometry starts with Gauss' *Disquisitiones generales circa superficies curvas* (1828). He decided that the most general way to describe a piece of surface was by imposing coordinates on the surface, such as latitude and longitude coordinates on the sphere. In this way, the x, y, and z coordinates of a point on the surface are expressed as functions of two variables, u and v. Now, arc length is expressed by a formula of the form

$$ds^2 = Edu^2 + 2F\,dudv + Gdv^2,$$

where E, F, and G involve the partial derivatives of x, y, and z with respect to u and v. Area is handled the same way. In the plane $dS = dxdy$. On the surface, dS works out to be $(EG - F^2)dudv$.

Gauss' great discovery in geometry is the concept of intrinsic curvature. Informally, surfaces look locally like bowls, or saddles, or pieces of a plane, and it is implausible that there are distance-preserving maps (known as isometries) from one to the other. For example, it is elementary that there is no isometry from the sphere to the plane. Triangles on the sphere are made up of arcs of great circles (these are the geodesics, or curves of shortest length, on the sphere), and the sum of the angles of a triangle on the sphere is always greater than 180°. An isometry of the sphere to a plane, were one to exist, would map a spherical triangle to a plane triangle whose sides were geodesics (and therefore straight), and the sum of the angles of such a triangle is necessarily 180°. So the angles are altered, but then the map cannot be an isometry. This contradiction shows that there is no isometry of the sphere to a plane.

Gauss defined the *curvature* of a surface at a point P by looking at how the perpendiculars to the surface vary from point to point (think of them as little telescopes pointing at the celestial sphere—Gauss was professionally an astronomer). He then looked at how the element of area at a point on the surface appears on the sphere; fairly flat pieces of surface all have perpendiculars that vary little in direction, but on tightly curved surfaces the perpendiculars vary a lot. More precisely, on this definition, the curvature of a sphere of radius R is $1/R^2$ at every point. The curvature of a plane is zero everywhere, as is the curvature of a cylinder at every point, and saddles have negative curvature.

A lengthy but routine calculation gave Gauss the curvature of a surface at a point as an expression in E, F, G and their derivatives with respect to u and v, but crucially not in a way that involves x, y, or z. Gauss deduced that the curvature is therefore independent of the particular embedding used to define the surface and would be the same for any surface isometric to the given surface. So curvature is an *intrinsic* quantity, and it can be found at any point once it is known how to determine lengths and areas on the surface as functions of the coordinates, which is the information provided by E, F, and G. Nor is it necessary that some higher-dimensional space exists in which the surface is curved. Gauss called this remarkable result the *Theorema Egregium* or exceptional theorem.

In the 1850s, Bernhard Riemann was the first to see that the non-Euclidean geometry of Bolyai and Lobachevsky was the intrinsic geometry of a surface of constant negative curvature. This exists even though, as was shown much later, there is no isometry of all of such a surface onto a surface in our customary three-dimensional space. After Riemann's work, and the similar but largely independent discovery of the Italian mathematician Eugenio Beltrami, were published in the late 1860s, non-Euclidean geometry became widely accepted among mathematicians, although philosophers continued to have doubts for many years.

With the acceptance that the work of Bolyai and Lobachevsky described a new geometry that could be the geometry of physical space came another important change in attitudes to mathematics. Henceforth, no account of

FIGURE 3.3. All the triangles in this figure are congruent in the sense of non-Euclidean geometry, and all the arcs are geodesics. Wikimedia Commons.

geometry could be true on logical grounds alone, as had often been people's opinion of Euclidean geometry. Any geometrical account of space would henceforth have to rest on some empirical measurement.

Riemann also took Gauss' ideas and developed them in any number of dimensions, showing that the study of intrinsic features of a surface or higher-dimensional "manifold" was possible. In particular he generalized the idea of intrinsic curvature to an n-dimensional setting. This further advanced the study of differential geometry, but what gave the subject its greatest impact was Albert Einstein's realization around 1915 that gravity could be understood as the intrinsic curvature of a four-dimensional space-time. He could do this, despite his initial distrust of advanced mathematics, which he felt often obscured more that it illuminated, because a number of mathematicians had by then worked through the sometimes formidable technical difficulties. In his endeavors he was joined by Tullio Levi-Civita in Italy and Hermann Weyl in Germany (see Scholz 2001). Each of these men elucidated what it means for a curve in a curved space to be a geodesic, the analogue of a straight line in plane: they defined as a curve along which a particle moving at constant speed experiences no acceleration. The concept became essential in understanding how objects respond to a gravitational field. Weyl in fact rewrote the entire subject of differential geometry in the five editions of his book *Raum Zeit Materie* (*Space Time Matter*) to adapt it to the needs of contemporary physics and meet the high standards of rigor required in twentieth-century mathematics.

DEVELOPMENTS FROM 1850 TO 1914

It would be quite wrong to say that after this explosion of new ideas before 1850 the rest of the century was devoted simply to making them work. In every topic mathematicians not only drew out the full significance of what had been done but went on to add ideas of equal or even greater importance. Nonetheless, for reasons of space, and to cope with the problem of rising difficulty, it will be helpful merely to note some of the achievements that built directly on the earlier material before turning to some inventions more characteristic of the later decades.

In analysis the ideas of Cauchy were steadily accepted and refined. His ideas of limits were gradually replaced with a more rigorous use of inequalities associated with Weierstrass—the epsilon—delta–*N* approach of modern analysis. Riemann took the approach of Dirichlet, the mathematician he learned from the most, and explored the world of the merely continuous functions. These functions differ greatly from differentiable functions and show how much had to be done once Cauchy's style of rigor displaced the naivety of Fourier's mathematical worldview.

Especially in the study of partial differential equations it became clear that equations with analytic coefficients and analytic initial conditions are very different from those with more general (say, merely differentiable) coefficients. Cauchy had given a proof in 1841 that every partial differential equation has a solution, and its one major flaw was identified by Sonya Kovalevskaya in 1871 and removed (see Cooke 1984), but only at the end of the century was it appreciated that the proof worked only in the analytic case; there is still no proof of this result in the general case.

Number theory blossomed, especially in Germany, and in Dedekind's hands became more abstract and structural. Dedekind made a very thorough study of the arithmetic and algebra of what are called number fields. Informally, a number field is the collection of all polynomial expressions in the rational numbers and the roots of a polynomial equation with rational coefficients. (For example, the collection of all expressions of the form $m+n\sqrt{2}$, where m and n are rational numbers, is a number field.) His work was of exceptional importance in the creation of modern structural algebra by the start of the twentieth century. So much so that Emmy Noether, the great creator of modern structural algebra, used to say that it was all in Dedekind. His work virtually called for the introduction of the ideas of Galois theory, and when David Hilbert wrote his definitive account of the subject (the *Zahlbericht*, 1896) Galois theory was a prominent theme.

Galois theory was written up by Camille Jordan as part of an independent theory of groups, so that it became possible to see what Galois had written so obscurely as fragments of a larger, cohesive work. This was a view energetically promoted by Felix Klein in Germany, who had learned group theory from Jordan in Paris until his studies were interrupted by the Franco-Prussian War in 1870–1, and Klein also emphasized the importance of group theory in the study of geometry. Group theory also developed as a subject in its own right, as a study of symmetries and configurations, and as the way to classify and analyze transformations in geometry.

If these achievements can be seen as building on what went before more or less directly, the second half of the nineteenth century saw work that was perhaps more surprising. Here there is space for only four of these: the origins of topology, some paradoxical aspects of real analysis, Hilbert's axiomatization of mathematics, and Georg Cantor's discovery of infinite sets.

The origins of what is sometimes called algebraic topology lie in several places, but one is in the work of Riemann on algebraic curves—curves defined by a polynomial equation in two complex variables. We have not been able to discuss the emergence of the theory of complex-valued functions of a complex variable, which was one of the great successes of the century and which touched many other branches of mathematics. It rests on Riemann's understanding of the implications of the definition of what it is for such a function to be differentiable at a point, and that definition is solely concerned with what happens in a neighborhood of that point. This gave Riemann the opportunity to explore functions defined on all manner of domains made up of overlapping small pieces of the complex plane, and this includes all the closed and orientable surfaces. Because one of these arises every time one has a polynomial equation in two complex variables this gave him an insight into the complex curves (or real surfaces) that these equations define. So, it is important to know what these could look like, and he gave a powerful argument to the correct conclusion that any such surface would be precisely one of the ones in this list: the sphere, the torus, the two-holed torus, or the n-holed torus for some $n > 2$.

This classification ignored all the details of how such a surface might look, say when given by an equation, and concentrated on the idea that there is a one-to-one, onto, continuous map from the surface to precisely one of the ones in the list. This is a truly fundamental distinction between them, and Riemann was able to show that it determined a lot about the kinds of complex functions that can be defined on the surface. Today it is called a topological classification, because it depends only on properties of a surface that survive under any continuous, invertible, transformation.

In the early 1880s, both Klein and the French mathematician Henri Poincaré, who contributed to almost every branch of the mathematics of his time, claimed that the geometry of any (closed, orientable) surface is locally the same as the geometry on exactly one of these three surfaces: a sphere, a plane, or a disc with a metric of constant negative curvature. This united the geometry of such surfaces with their topological classification but the first proofs of the claim were not found until 1907.

Paradoxical figures in analysis arose in another type of topological argument concerning such questions as what can be the domain of a function of a certain kind. Intuition strongly suggests that a closed curve in the plane that never crosses itself has an inside and an outside, but is that always true? And are the regions, if they exist, always each in one piece, or might they come in several? One reason this matters is that a thorough mathematical theory of gravity, magnetism and electricity requires that it should be possible to define what is called a potential function for the interior of any such curve. Of course, any applicable theory requires that these ideas also apply to surfaces in space: do closed surfaces with no self-intersections have a single interior and exterior?

To answer the first set of questions, we have to define what a curve in the plane is in such a way that it can be studied with the tools of analysis after Cauchy. The simplest will be the curves that are the graphs of a function f. These are the sets of points in the plane with coordinates $(x, f(x))$ for some range of real numbers x. Around 1830, the Czech priest, philosopher, and mathematician Bernhard Bolzano was the first to define a function that is continuous but not differentiable at any point (Russ 2004). Intuitively, its graph is very crinkly, so much so that it never has a tangent anywhere. There is no natural way of excluding them from mainstream mathematics; Poincaré showed early in the 1880s that they turn up naturally, so they have to be looked at. And in the early twentieth century the theory of Brownian motion was shown to depend on continuous, nowhere differentiable paths in space, so they are important for physics, too.

Does every curve have a length? From the formula $ds^2 = dx^2 + dy^2$ it might seem obvious that such non-differentiable curves cannot have a length, but might there not be a better way of defining length that finesses this difficulty? The von Koch curve, which fits in a bounded region of the plane, shows that there is not. It is defined by a sequence of approximations, and at every stage the approximating curve has a length 4/3 times the preceding one, so the length of the limiting curve must exceed any finite amount.

With such curves in mind, let us ask again if a closed curve in the plane that never crosses itself has an inside and an outside? In fact, intuition holds up, and the statement is true. It was first proved by Jordan in the 1890s, but his proof was thought to be doubtful and was replaced by others, until it was re-examined recently and found to be satisfactory. Moreover, the inside is always in one

FIGURE 3.4. The von Koch snowflake curve is continuous but nowhere differentiable. Photo by Encyclopaedia Britannica / UIG / Getty Images.

piece: any two points in the interior can be joined by a curve that lies entirely in the interior. The same is true of the exterior, too. Topologically, the interior is the same as a disc, and the exterior is the same as a disc with a point removed. All these statements are true for the analogous set-up in three dimensions, except the one about the exterior, which need not be topologically equivalent to a ball with a point removed.

Surely it is obvious that a curve has dimension 1 and a plane dimension 2? Cantor was the first to find a one-to-one map of an interval to a square, but his map fails terribly to be continuous. But in 1891, the Italian mathematician Giuseppe Peano found a continuous map of an interval onto a square—the curve passes through every point of the square. It was some time before it was shown that there is no continuous *invertible* map of an interval to a square, and the concept of dimension was rescued.

Hilbert's axiomatization of mathematics grew out of an emerging consensus in the 1880s that Euclid's *Elements* was not rigorous by contemporary standards and could not be patched up. Instead, it would be necessary to start from scratch if an axiomatic account of elementary geometry was to be given. Italian mathematicians took up the cause, inspired by Peano's high standards of rigor, but they were unambitious and aimed only to provide a rigorous account of geometry for schoolteachers. A little later Hilbert drew on a contemporary German debate and addressed the same question, to the surprise of those who knew him, who could see no prospect of new results in the subject comparable with what Hilbert had achieved elsewhere.

Hilbert succeeded in founding elementary geometry on five families of axioms, but he did much more. He investigated various sub-collections of these axioms, and axiom systems with other axioms. He asked himself whether anything existed that corresponded to any given set of axioms. How could one be sure that the axioms did not imply a contradiction? Axioms for Euclidean geometry could be justified by pointing to the usual coordinate geometry of the plane, and Hilbert often answered his question by showing that a collection of axioms was satisfied by a coordinate model. This led in the twentieth century to the development of model theory in mathematical logic.

He found that some axiom systems had coordinates that were not necessarily real numbers but still satisfied some basic properties. For example, Pappus' theorem required coordinates that obeyed the associative and commutative laws of multiplication, but Desargues' theorem in the plane only required the associative law. So Pappus' theorem implied Desargues', but not the other way round. As this example shows, Hilbert investigated what systems of axioms are or are not compatible with other systems, and his former professor Adolf Hurwitz, a distinguished analyst and number theorist, was the first to praise him for creating a new branch of mathematics, that of axiom systems. It was to prove one of the most productive ways of organizing and advancing algebra in the twentieth century.

Cantor's theory of infinite sets was part of a response to concern about the biggest single gap in Cauchy's presentation of real analysis: its use of variable quantities. It proved more rigorous to talk about real numbers—but what are they? What is required to take the rational numbers and fill in all the gaps between them? That is, to define all the irrational numbers, and to prove that set of all rational and irrational numbers has no gaps? Several mathematicians gave answers to this question; the two that have survived are a topological one due to Cantor and one by Dedekind that uses the fact that the rational numbers are ordered.

Cantor had begun his mathematical career by investigating the sorts of functions for which Fourier series may fail to give an accurate picture, and he asked if such a representation was necessarily unique. To answer it he began to look at sets of points in the plane where a function could behave badly. He said that a point *P* was an accumulation point of a set of points if every neighborhood of *P* contains a point of the set other than *P*. Obviously only infinite sets can have accumulation points, so Cantor began to classify sets into types. A finite set is of type 0. An infinite set is of type 1 if the set of its accumulation points is finite (like the set of points $1/n$, where n is a positive integer, which has zero as its only accumulation point) A set is of type 2 if the set of its accumulation points is of type 1, and so on.

He now switched his attention from the sets to the idea of types, and began to explore the idea of succession, and the order in which elements in a set might occur. He found that the same set can be ordered in many ways. For example, the positive integers are usually presented in this order:

$$\{1, 2, 3, 4, \ldots\},$$

but they could equally well be given in this very different order:

$$\{2, 4, 6, 8, \ldots 1, 3, 5, 7 \ldots\},$$

in which the even numbers fill up the first infinity, the odd numbers the second one. The first ordering is of order type w, the second one of type $w + w$. All of these sets are well-ordered. This is a tricky concept. A set is well-ordered if it is totally ordered (if a and b are in the set then either $a < b$ or $a = b$ or $b < a$) and every subset of the set has a least element. Cantor believed this was true of all sets, but it not at all clear how the set of all real numbers can be well-ordered, and this question was to become one of Hilbert's famous Paris problems.

Inspired by this insight, Cantor began to look at the sheer size of some sets. He said a set was *countable* if it could be put into a one-to-one correspondence with the positive integers. He could show that the set of rational numbers is countable, and so is the set of all algebraic numbers (the real or complex numbers that are roots of polynomial equations with integer coefficients). But he could also show that the set of all real numbers is not countable, which means that there are numbers that are not algebraic. In fact, his result means that most real numbers are not algebraic.

Could there be a set larger than the set of all rational numbers but not as large as the set of all real numbers? Cantor thought not, but he could never prove it. When the problem was taken up with much more rigorous tools the conclusion was that Cantor's claim cannot be proved from the usual foundations of set theory.

Mathematicians divided over the merits, and even the validity, of Cantor's work for some time. The result that a set S cannot be put into a one-to-one correspondence with the set of all its subsets, $P(S)$, and so must be said to be smaller was a particular worry, because it immediately implies that there are sets of arbitrarily large size (such as S, $P(S)$, $P(P(S))$, . . .). It seemed to Leopold Kronecker, Poincaré, and others that the human ability to reason about such huge sets, whose members cannot possibly be examined one by one, was not certain and so such sets should not be used in mathematics.

The end of the century and the start of the next was an optimistic time for Europeans, and many plans were made to celebrate the transition, none more so than in Paris, where a six-month long World's Fair was organized to mark the event. To underline the centrality of French contributions to every walk of scientific life, three week-long Congresses were organized on Philosophy, Physics, and Mathematics. Poincaré spoke at all of them, and interesting, even lasting, though his contributions proved to be, in retrospect the star of the mathematical show was Hilbert (Gray 2000).

By 1900, Hilbert was the leading mathematician in Germany, and Poincaré's only peer. He had made his name in three separate branches of mathematics, and was turning to a fourth. First had come a series of ground-breaking papers in algebra, then number theory, and then the study of axiom systems. Now, his friend Hermann Minkowski advised him to seize the moment and predict the future, and this is what he did. As he put it, at the start of his talk, "Who among us would not be glad to lift the veil behind which he future lies hidden; to cast a glance at the next advances of our science and at the secrets of its development during future centuries?"

After a lengthy discussion on the importance of problems in mathematics, Hilbert presented ten problems in his talk, and a total of twenty-three in the published proceedings. They caused little stir at first, but gradually their reputation grew until it was said that if you solved one of them you joined the Honors Class of Mathematicians. This is a measure of how shrewdly many of them were chosen, and of the influence of Göttingen, where Hilbert presided, up until its demise in 1933 under the Nazi government.

Hilbert drew his problems from many areas of mathematics. First came problems in the foundations of mathematics: Cantor's problem of establishing the cardinal number of the continuum, and the problem of showing that the axioms for arithmetic are consistent. Hilbert asked Ernst Zermelo to try to axiomatize the study of sets, which would make them amenable to rigorous study, and Zermelo published his account in 1908. Numerous French mathematicians objected, most vociferously to the so-called *axiom of choice*, which allows one to choose an element from each set in an infinite family of non-empty sets (and implies that every set can be well-ordered, as Zermelo had already shown in 1904). Amusingly, it transpired that these mathematicians had been confidently doing exactly that in their own work elsewhere in mathematics (Moore 1982). However, this did not invalidate the worry that infinite sets might be a source of error, and the debate continued through the twentieth century, driven by the opinions of L. E. J. Brouwer (van Dalen 2013) and Hermann Weyl (see Chapter 2 in Volume 6).

If the cardinality of the continuum was of recent vintage, the second is surely strange: we have been doing arithmetic for millennia, what could go wrong? But Hilbert had relied on the consistency of arithmetic in his work on the axioms for geometry, and hoped to derive mathematics from set theory, so this problem asked to bridge the gap between geometry and set theory. Decades later, the cardinality of the continuum was shown to be independent of the usual axioms for set theory, and any theory of arithmetic that included multiplication fell to Gödel's discovery of the incompleteness theorems in the early 1930s (which says, loosely that in any axiomatization of mathematics there will always be statements that are true but not provable, so arithmetic is not provably consistent).

Among the other problems were some drawn from number theory. One was the famous Riemann hypothesis, which has drawn an immense amount of work and is still unsolved at the time of writing. Another asked for an algorithm capable of determining whether any given problem about polynomial equations in integers and having integer coefficients is solvable in integers (the so-called Diophantine problems, after the Greek number theorist Diophantus). This was shown to be impossible only in the 1970s. Towards the end, Hilbert turned to problems about partial differential equations, the calculus of variations, and one that acknowledged the vitality of an early contribution of Poincaré.

This selection of the 23 problems highlights the ones that proved exceptionally difficult and important. Others could never be solved (axiomatize physics), but many were, and of course Hilbert did not succeed in every case in picking topics that would shape the twentieth century in mathematics. Nor should it surprise anyone that major topics sprang up that he had not discussed. But what was perhaps more lasting was his creed that problems matter more than, say, theory, or rather that the test of a good theory is its ability to solve hitherto unsolved problems.

To that end, Hilbert chose two problems from the seventeenth century: the brachistochrone problem, which asks for the curve of quickest descent joining two points, and Fermat's last theorem. He indicated the many advances that had come from attempts on one or other of these problems, even if they were unsuccessful. The question of the brachistochrone not only indicates the richness to be found in applying mathematics to physics, it is the archetypal problem in the calculus of variations that still stood in need of a rigorous theory. Here Hilbert took the opportunity to stress that rigor in mathematics did not make it harder, by cutting the researcher off from intuition, but made it simpler. Nor were intuition and experience to be barred, but rigor forces the endeavor to find simpler methods of proof, and more fertile arguments. As for Fermat's last theorem, number theory in the hands of Gauss, Kummer, Dedekind, and Kronecker had led the way to the creation of the modern theory of algebra, and Hilbert surely saw that those developments had a long way to go.

In line with these general remarks, Hilbert also reflected on the seemingly abstract nature of mathematics, and how problems could best be tackled, say by first ensuring that simpler versions of the unsolved problem had been fully worked out. The idea here is that solution methods to simpler versions of the problem at hand may contain valuable hints, which become more visible when an abstract, clean-cut version of the problem is at hand.

Famously, Hilbert proclaimed his belief that there was no unsolvable problem in mathematics. He of course accepted that, for example, Galois had shown that most polynomial equations of degree 5 or more cannot be solved by radicals, but suggested that in such cases the task shifted to finding other solution methods. This vision is generally held to have faltered with Gödel's theorems, and still more with the negative solution to the problem about finding an algorithm for every Diophantine problem. But it is not merely an example of European optimism, it is probably, with suitable caveats, the belief of every mathematician who has done important, lasting work, that somewhere in the obscurity of their research there is an answer, a light switch such as Andrew Wiles spoke of when he finally solved Fermat's last theorem, that illuminates the room they are in before they plunge into the darkness of the next one.

CONCLUSION

Some of the surprising results just discussed make it clear that a high degree of rigor in the proofs would be required to make the conclusions acceptable. In fact, what all these results demonstrate is that naive intuition was proving incapable of asking the best questions and of finding the sharpest means to answer them. The German mathematician Felix Klein distinguished between what he called naive and refined intuition. The former had to be replaced, in his view, by the latter in the education of a mathematician. He gave as an example the unexpected nature of curves, and remarked that naive intuition deals not with a mathematical curve but with something like a very thin strip.

The move away from naive intuition and towards precisely defined, if abstract, concepts, was necessarily accompanied by the development of new types of proof, in which every appeal to intuition was to be replaced by reference to a definition or an already established theorem. Of course, this process was carried out imperfectly. Two of the most important mathematicians of the century, Riemann and Poincaré, were regarded as both profound and exceedingly difficult because their insights often considerably exceeded their ability to supply proofs. But it was almost always advantageous for later mathematicians to create the right framework for their insights to be given rigorous proofs.

This essay has only been able to sample some of the many inventions in mathematics in the nineteenth century. Even so, it is clear that a rigorous account of the calculus was finally provided, that novel geometries were invented and Euclidean geometry lost its central role, that the new subject of topology was beginning to prove its worth, and that elaborate mathematical structures such as groups had begun to fill a new field of abstract algebra. Even the idea of what it is for a mathematical object to exist was changing: it might be an idealization of a physical process, but it had to be expressed in a way that was free from contradiction. Mathematics deepened people's insight into many aspects of the natural world, but it also left the natural world behind in its search for things to describe. How these two tendencies interacted after 1914 is one of the themes taken up in Volume 6.

CHAPTER FOUR

Mathematics and Worldviews

Enlightenment Ideals and Mathematical Modernism

IVAHN SMADJA

In a speech delivered before the National Convention on the 5th of Brumaire, an III (1794), Joseph Lakanal eloquently defined the new role French Revolutionaries envisioned for mathematics. Echoing Jean d'Alembert and Nicolas de Condorcet, he glorified mathematical analysis as a scientific ideal aimed at promoting the transparency of reason. What is more, he claimed it served as a decisive political instrument for the diffusion of Enlightenment:

> Analysis must equally enlighten all sciences and become the universal tool of all human knowledge. While political freedom as well as freedom of industry and trade will destroy the monstrous inequalities of wealth, analysis, applied to all kinds of ideas, will destroy in each and every school the inequality of Enlightenment, [which is] even more fatal and humiliating. Analysis is an indispensable instrument in any great democracy.[1]

In this context, mathematics took on a privileged status as a model of rationality and as a new basis for education. Within a decade, the French Republic created new teaching institutions in which the lion's share of the curriculum went to mathematics. While the École polytechnique was to supply the nation with engineers, the École normale de l'an III would be devoted to educating teachers, who in return were to propagate exactness and clarity throughout the entire social body (see Chapter 2). Such major figures as Lagrange, Monge and Laplace took an active part in this process, which reshaped the teaching of mathematics by presenting general methods in a simple and most lucid way that made it accessible to the largest possible number of people.

An emblematic expression of the liberating value then attached to mathematics is provided by Stendhal's account of his own youthful passion for mathematics which led him to prepare the examination for Polytechnique (Hodgkin 1981). "My enthusiasm for mathematics, he confessed, may have had as its principal basis my loathing for hypocrisy . . . In my view, hypocrisy was impossible in mathematics." At a loss to explain why a negative multiplied by a negative equals a positive number, his provincial professor of mathematics finally exclaimed in exasperation, "But it's the custom!" The École polytechnique attracted young talents in the 1790s precisely because it promoted reason over custom. It held the Enlightenment promise of imparting true knowledge and understanding in a unifying, progressive and emancipatory way. Stendhal's path, however, took another direction. Arriving in Paris the day after the 18th of Brumaire, he enlisted in a regiment of dragons on its way to Italy.

After Napoleon's rise to power, and even more so after the Restoration, the "exemplary view" of mathematics receded in favor of a "separatist," rigorous definition of the subject, championed by Augustin-Louis Cauchy (Richards 1991). His "Introduction" to the *Cours d'Analyse* (1821)[2] verged on sounding like a pamphlet railing against Enlightenment universalism and its mathematical egalitarian ideology. Abstract rigor prevailing over conceptual clarity had epistemological import but it also carried strong political overtones. Mathematics was no longer to be seen as a vehicle for laymen aspiring to become citizens. It was a stern self-insulated discipline, structured by abstract definitions and intended for committed specialists only. It offered no prospective meaning outside its domain of jurisdiction, staked out strictly by dint of abstract stipulations. "Let us not imagine," Cauchy warned, "that we can attack history with formulas nor that morals can be sanctioned by theorems of algebra or of integral calculus" (1821: viii). Even though his work built on Lagrange's algebraic foundation of the calculus using Taylor series expansions, the shift in emphasis was marked (see Chapter 3).

Lagrange's attempt to reduce the calculus to algebra was consistent with the Enlightenment conception of analysis. His goal was to rise to a privileged unified standpoint free from the hodge-podge of rival imperfect methods ranging from infinitesimals and evanescent quantities to Leibnizian differentials and Newtonian first and last ratios. Cauchy, by contrast, "tried to give the methods all the rigor one demands of geometry, in such a way as never to fall back on reasons taken from the generality of algebra" (1821: ii). The heretofore-entrenched tendency to "grant algebraic formulas an indefinite scope," he argued, was not just mathematically misguided; it was morally and politically disastrous. In Cauchy's view, it was "a serious mistake" to think, as the French *philosophes* did, that certainty was only to be found either in mathematics or in the testimony of the senses. "There [*were*] truths other than those of algebra, realities other than those of sensible objects," as historical facts and "a host of questions in religion, morals, politics" (1821: iii–vii) allegedly testified.

In Restoration France, Cauchy's mathematics was institutionally central (Belhoste 1991), only counterbalanced by Monge's school, whose fruitful view of geometry as founded on intuitive and experiential knowledge made it a powerful undercurrent, reminiscent of Enlightenment universalism (Daston 1986). Abroad, the British analysts largely ignored it, while holding instead to the opposite late eighteenth-century unified view. On the continent, by contrast, Cauchy's epsilon-delta methods were soon adopted and systematized from Abel to Weierstrass, while the arithmetical continuum of real numbers underlying them was eventually rigorously defined by Dedekind and Cantor (Petri and Schappacher 2007). By the time these leading figures gradually consecrated Cauchian (or Weierstrassian) rigor as the nineteenth-century mathematical standard, the earlier clashing moral and political worldviews had long ago faded away. On the path to *fin-de-siècle* mathematical modernism, however, Enlightenment ideals, whether valued or repelled, remained a standard reference point for practitioners and interpreters of mathematics trying to find their bearings in modern times. From Comtean positivism and English utilitarianism, through to German mid-century scientific cosmopolitanism and Mach's "historical Enlightenment" (Uebel 2021), these reverberating ideals would be reinvented and contradictorily interpreted through the distorting lens of historical distance. Defining the present with reference to the past involved complexities arising from Enlightenment's legacy that were inextricably continuous and disruptive.

Cultural historians have disagreed sharply as to whether and, as the case may be, how and when the impact of the Enlightenment ended. Major confluent analyses, aptly labeled the "Koselleck-Foucault thesis," strongly advocate for discontinuity. They claim that a fundamental breach occurred at the turn of the eighteenth to the nineteenth century, reflected in the role of time, as transparently unfolding chronologies gave way to germinating "inner histories" (Jardine 1999). From the actualization of preexisting possibilities, registered in advance in tabular form, one shifted to a new notion of time in which genuine novelty would arise, whether in the realm of living things, history or culture (Foucault [1966] 2002; Koselleck [1979] 2004). In the entry, "Élémens des sciences" of the *Encyclopédie*, d'Alembert still considered that, if knowledge were entirely reduced to its principles, the human mind could grasp it, "as the supreme intelligence, from one indivisible viewpoint . . . with this difference, however, between God & man, that, placed at this viewpoint, God would see everything at a glance, whereas man needs to proceed stepwise, one thing after another, so as to gain knowledge." Mathematics therefore was but a paradigmatic expression of the "*esprit humain*" gradually unfolding its inherent features in a progressive historical narrative (Richards 2006). When the Enlightenment ended, a new sense of historicity would thus emerge.

Other interpretations, however, argued for a measure of continuity, stressing how Enlightenment's legacy kept shaping our modern world. Max Horkheimer and Theodor Adorno claimed that "Enlightenment's program was the disenchantment of the world" (Horkheimer and Adorno [1969] 2002). They contended that the process was still ongoing and that mathematics was central to it. Max Weber, who coined this notion of disenchantment, had pictured the Enlightenment as the "laughing heir" of the Protestant work ethic central to capitalism and an ever-increasing rationalization of social life. Modernity left individuals trapped in the "iron cage" of technical rationality, calculation and control. Horkheimer and Adorno even held that the core of the entire historical process lay in the way the Enlightenment enthroned mathematics as its privileged self-mirroring instrument, while discarding anything that failed to conform to the "standard of calculability and utility." From the outset, they argued, the Enlightenment was doomed to fail as an emancipatory project; its legacy revealed itself to be oppressive. This involved a deep-seated paradox, born of an imbalance between two distinct drives. Instead of entailing a proportionate growth of autonomy for the individuals, as was Enlightenment's great hope, the growth of technical capabilities resulted in an intensification of power relations (Foucault 1984). While Horkheimer and Adorno only intimated that power and knowledge were synonymous, Michel Foucault documented how various Enlightenment-era schemes for producing quantified knowledge eventually brought about a world dominated by disciplinary order, surveillance and social control (Foucault [1975] 2012).

Horkheimer and Adorno's point, however, was more radical still. For not only was the world conceived in advance in terms of mathematics as regimented and preemptively knowable, even nature itself was regarded as a mathematical manifold. However, this kind of knowledge presumably came at a price. Subjugating the world, both natural and social, to mathematical apparatus only results in holding humankind captive to irrational forces different from, but just as powerful as those that prevailed in the enchanted world of myths and superstition. Gradually realizing that the world they shaped for themselves by dint of mathematics and technology admits of no way out, human beings would end up "expect[*ing*] it to be set ablaze by a totality which they themselves are and over which they are powerless" (Horkheimer and Adorno [1969] 2002: 22). These words, written when the threat of nuclear extinction was on everyone's mind, retain some potency even today in an age of major ecological crisis. Such concerns, however, were already rooted in *fin-de-siècle* debates that echoed tensions between mathematical modernists and counter-modernists

(Mehrtens 1990), a conflict that reverberated well into the twentieth century (see Chapter 4 in Volume 6).

Taking his cue from a lecture given by Max Planck in Leiden in 1908 (see the concluding section below), Heidegger modeled his analysis of the essence of modernity upon a conceptual clarification of the ontological meaning of modern "mathematical physics." The fundamental presupposition that "something should be specified in advance as that which is already known"[3] implied that "every natural event must be viewed in such a way as to fit into this ground-plan of nature" (Heidegger [1938] 2002). In his view, the one "world-picture" [*Weltbild*], coextensive with the modern age, thus denoted the all-encompassing metaphysical structure underlying modernity as a whole, which left room in return for several competing "worldviews" [*Weltanschauugen*] corresponding to the different ways to live out this modern existence. Horkheimer and Adorno's interpretative categories, as well as Heidegger's, carry their load of angst, harking back to what Husserl diagnosed as the crisis of the European sciences, and exacerbated by the experience of twentieth-century totalitarianisms and the Cold War. Even so, something of their meaning remains with us. This chapter proposes to analyze the complex ways in which, over the stretch of time extending from the aftermath of the French Revolution to the Great War, mathematics in flux interwove with worldviews in collision.

THE LAPLACIAN WORLDVIEW AND ITS OPPONENTS

Under the Consulate and throughout the Napoleonic era, the *savants* came to exercise unprecedented influence in the State (Fox 2012). Among them, the mathematician Pierre-Simon Laplace dedicated a volume of his *Mécanique céleste* to Bonaparte, became Senator, and dominated French science in the period. With the chemist Claude-Louis Berthollet, he presided over the so-called *Société d'Arcueil* which, partaking of a French salon and "a research school for neo-Newtonian science," gathered many young promising scientists such as Jean-Baptiste Biot and François Arago (Crosland 1967). The Laplacian worldview aimed at extending the Newtonian regularity of nature from the astronomical to the molecular scale. It sought to account for all phenomena in terms of central forces, either attractive or repulsive, which, by analogy with universal gravitation ruling celestial phenomena, were conceived by extension as "mechanical" (see Chapter 4 of Volume 4). At the molecular level, one thus assumed that short-range forces, not necessarily obeying a simple inverse-square law, could account for optical refraction, capillary action, crystal properties, or even chemical affinities. Laplacian physics presumed these forces applied not only to particles of ponderable matter but also to imponderable fluids such as heat, light, electricity and magnetism. While one imagined these weightless fluids to be causally independent of one another, each consisted of

FIGURE 4.1. Pierre-Simon Laplace (1749–1827). Colored lithograph from a painting by Sophie Feytaud, France, approx. 1841. Photo by Fototeca Gilardi / Getty Images.

mutually repulsive particles, which in turn were attractable by ponderable matter (Fox 1974).

A revolt against this Laplacian worldview occurred already within the following decade 1815–25. The first serious challenges came from Joseph

Fourier's analytical theory of heat (see Chapter 5) alongside Augustin Fresnel's critique of the corpuscular theory of light (Buchwald 1989; Darrigol 2012). Trouble also came from abroad, especially with Hans Christian Oersted's discovery of electromagnetic induction. François Arago, a former regular attendant of the Arcueil meetings, soon emerged as a voice of support for the anti-Laplacians. Taking over the so-called "course of machines," conceived by Gaspard Monge and created by Jean Nicolas Pierre Hachette at the École polytechnique, Arago praised the steam engine and the railroad as ultimately beneficial to the working classes, hence as a vector of political progress and republicanism. Through attempts at measuring and comparing the productive forces transformed and converted in all sorts of machines, both classical and steam, the mathematical concept of mechanical work, defined as force times distance in the direction of the force, gradually emerged during the Restoration. While Claude-Louis Navier stressed the need "to establish a kind of *mechanical currency* . . . with which one may estimate the quantities of work employed in order to effect all kinds of manufacture" (Grattan-Guinness 1984), Gaspard Gustave de Coriolis took into account friction and resistance. Now one could conceptually separate "useful work" from waste, a distinction that was to be later elaborated upon across the Channel (Wise and Smith 1989–90, 1990). Therefrom would eventually result a shift from a world of stability and equilibrium modelled upon the balance, the lever and the clock to a world of irreversible processes brought about by the physics and the economy of the steam engine.

In 1820, the Dane Oersted stumbled upon the interaction between electricity and magnetism, an unexpected phenomenon that violated central tenets of Laplacian orthodoxy. Not only did this reveal fluids of different types acting upon one another, but also the action itself was apparently at odds with the central forces of Laplacian physics. Since the magnetic needle pointed in a direction deviating from that of the conducting wire, this suggested that the forces involved were rotational. Oersted's discovery almost immediately aroused the interest of French scientists. While Arago reported on it to the Paris Académie des sciences, André-Marie Ampère launched into reproducing and modifying Oersted's experiment, which he envisaged differently (Steinle 2016). Thus began a new program for investigations into the "unity of nature," which sought to find underlying connections between the phenomena of electricity, magnetism, light and heat, in both animate as well as inanimate objects. This goal was pursued, although in different ways, by proponents of German *Naturphilosophie* as well as French ether physics (Caneva 1997). Oersted had drawn inspiration from Schelling's philosophy of the so-called "absolute identity," which claimed that the dialectical processes empirically displayed by nature were none other than the inner dialectic of the mind made visible. Oersted's joint work with Johann Ritter on galvanism had thus followed Schelling's writings on the essence of polarity.

Ampère, however, did not approach Oersted's finding along these lines. Imbued with Cartesian dualism and Catholicism, he maintained that matter and mind were two essentially different, although interacting, kinds of things. Together with Fresnel, Arago and many of the anti-Laplacians, Ampère held a worldview in which electro-magnetic phenomena, heat, light and even chemical interactions were conceived as vibrations of an all-pervading ether (Caneva 1980). Following Lamarck's work, however, one viewed these imponderables as extending beyond lifeless nature to the "milieu" of the living realm, in which modifications of the ether could account for transformations occurring in organisms (Canguilhem [1965] 2008). Under the influence of the Catholic philosopher Maine de Biran, whom he enjoined in return to read Kant, Ampère embraced the notion that knowledge of both the self and the material world was to be jointly rooted in, and originally gained through, the inner sense of muscular resistance that one feels when moving one's own body (Tresch 2012). Similarly, knowledge of the external world implied an exploratory process seeking to identify which effects result from varying circumstances. Ampère thus experimented with partly movable frames, voltaic conductors, and coiled wires, in an effort to make mathematical laws perspicuous. In compliance with the norms long prevailing in French physics, Ampère's paradoxical recourse to Laplacian rhetoric at the very same time he subverted it has been labelled as a paradigmatic case of "double discourse" (Blondel 1982).

Change also came from another side. From the 1820s onwards, most European countries began systematically to publish social statistics. The remarkable stability of mean values in crime or suicide rates suggested that statistical regularities revealed some structural feature of the social order. The Belgian statistician Adolphe Quetelet considered that, although mere abstractions, these statistical averages had a deeper reality than the mismatch of any collection of actual individuals, so that social statistics were the appropriate tool for guiding social reforms (see Chapter 5). By contrast, Auguste Comte strongly opposed this use of statistics, which in his view amalgamated together social factors that belonged to different stages of development. Moreover, he denounced as chimerical the attempt to render social studies positive by the shortcut consisting in subordinating them to a presumably illusory mathematical theory of chances. Balzac may have sensed these tensions when, in his novel *Le Curé de village* (1841), he wrote, "society isolates everyone, the better to dominate them, divides everything up to weaken it. It reigns over the units, over numerical figures agglomerated like grains of wheat in a pile." Dickens' heroine in *Hard Times* (1854) also protested against the supposedly philanthropic utilitarian statistics, drowning individuals in numbers. As statistical laws gained objective significance in the social realm, chance gradually appeared as an autonomous feature of the world (Hacking 1983; Gigerenzer et al. 1989).

SAINT-SIMONIANISM AND POSITIVISM

A committed Saint-Simonian, Émile Clapeyron, who turned Sadi Carnot's heat cycles into mathematical diagrams (see Chapter 5), may have been receptive to the potent view of a world ruled by heat offered in the opening pages of the *Reflections on the motive power of fire*. Whether natural or harnessed by technology, heat was hailed as the all-pervasive cause of motion, from atmospheric phenomena to volcanic eruptions, and from mining, ship driving, and canal digging to steel forging, and cloth-weaving. Both Clapeyron and Carnot belonged to the same seething milieu during the aftermath of the Revolution of July in 1830. Sadi Carnot's younger brother, Hippolyte, was a regular member of the Saint-Simonian circle, which Clapeyron also attended.

Claude-Henri Rouvroy de Saint-Simon was an unclassifiable, though highly influential figure in the French intellectual scene of the first half of the century. A great-nephew of the chronicler of the court of Louis XIV, an aristocrat intent on setting a new political order for the post-Revolutionary era, and an adventurer who joined Lafayette to fight for the American Revolution, Saint-Simon diagnosed the break-up of the old social order and the moral crisis brought about by the Revolution. He advocated social reorganization and planning, as well as a grand shift of power—"spiritual power" being transferred from the Church to the *savants*, but also "temporal power," formerly feudal and military, to be taken over by the emerging "industrial class." Saint-Simon lived next to the École polytechnique for years, and he regularly invited Laplace, Monge and Berthollet to his dinner table; his doctrine was also akin to the technocratic ideals that inspired this institution. Saint-Simon, however, used the terms "industry" and "industrialism" idiosyncratically, giving them a distinctive meaning. His "industrial class" was in no way similar to the so-called "bourgeoisie," for it gathered all members of society engaging in productive work, from *savants* and artists to bankers and artisans, but also farmers and manufacturers.

In contrast to Karl Marx, Saint-Simon identified political agency, not the (capitalistic) mode of production, as the main driving force in history. In his view, the main obstacle standing in the way of "industrialization" and social progress was power in the hands of a noxious class of aristocrats, speechmakers and theorists. This segment of society held sway over the political arena and for them he reserved the term "bourgeois." Within the "industrial class," however, no class struggle divided laborers from captains of industry—a notion that became increasingly disputable as the nineteenth century grew older. This also accounts for why both liberals and socialists laid claim to Saint-Simon's ambiguous legacy. After the master's death in 1825, the Saint-Simonian school proved an active force in France as well as abroad by promoting industry in a peaceful and humane way in keeping with the emancipatory ideals of the French Revolution. In his *Système de le méditerranée* (1832), Michel Chevalier, an

FIGURE 4.2. Auguste Comte (1798–1857). Wikimedia Commons.

eminent Saint-Simonian and editor of the widely read newspaper *Le Globe*, praised the railroad as a "most perfect symbol of universal association." He advocated the extension of a global network of communications that would turn the Mediterranean into a unified system, enhancing the moral bonds of humanity between north and south.

In 1816, freshly graduated from the École polytechnique, Auguste Comte became secretary to Saint-Simon, thereby completing his mathematical training with an early immersion in the doctrine of the French social reformer. Although his first publications betray the influence of his mentor, Comte later distanced himself from the Saint-Simonians by refusing to endorse their paean to unrestrained industrial mastery over the whole world. His mature work bespeaks an acute awareness of humanity's inescapable dependence on its milieu, which meant setting limits on the promotion of power over nature in order to safeguard the earth as its enduring dwelling. By the late 1820s, Comte had presented a first sketch of his ideas in public lectures delivered at his home, in the presence of such eminent figures among the Parisian scientific elite as Fourier and Navier. The main tenet of Comtean positivism consisted in the rejection of metaphysics in favor of empirically based knowledge, which implied shifting the focus from ultimate causes to lawful relations between phenomena. Hence the new analytical theory of heat (see Chapter 5), relabeled "mathematical thermology," received pride of place, and Fourier was recurrently praised as an exemplar of this new positivistic attitude. In his *Cours de philosophie positive*, whose publication spread from 1830 to 1842, Comte offered a rational classification of the sciences and the principle of development governing them, his so-called law of three stages. In addition to claiming that every field of knowledge passed from the theological (or fictive) through the metaphysical (or abstract) to the positive (or scientific) stage, Comte also opposed reductionism. In his view, each science possessed its own proper concepts and methods, irreducible to those of other sciences. His motto "from science comes prediction; from prediction comes action" encapsulated the philosophical insight that human beings can only grasp the laws ruling natural phenomena, hence orient the course of things, owing to the artificial tools of theory. Accordingly, more concrete and special sciences should be subordinated to the more abstract and general ones. Therefore, mathematics comes first, then astronomy, physics, chemistry, physiology, biology and lastly sociology.

In January 1835, Comte presented a memoir at the Paris Académie des sciences entitled "*Cosmogonie positive,*" in which he proposed a mathematical justification of Laplace's nebular hypothesis. This assumed that the planetary system formed out of a rotating mass of hot gas whose condensation successively left behind outer rings. These in turn accreted into planets to which most of the angular momentum was transferred. Although Comte's classificatory framework precluded mathematics from encroaching on other sciences that dealt with phenomena of higher complexity, it nevertheless permitted a contribution of a more refined kind. Mathematics could provide biology or cosmology with idealized models, thereby endowing scientific hypotheses with sufficient precision to make them amenable to empirical control. In order to fill this gap, Comte packaged various mathematical formulae into one scheme loaded with the information required to allow for the unfolding of the intended cosmogonic narrative (Schweber 1991). Within French scientific circles, a refutation promptly came which pointed out a vicious circle. His detractors claimed this was no surprise: Comte could only extract as much from his formula as he had put into it in the first place. In Britain, by contrast, where the nebular hypothesis was a popular theme among Victorians who thrilled to discuss their conflicting evolutionary worldviews (Brush 1987; Schaffer 1989), Comte's memoir caused a sensation until the astronomer John Herschel authoritatively pronounced against it in 1845.

In Comte's view, the sciences differed in terms of the way they marked time. This meant treating different orders of phenomena as regulated by different interconnected patterns of time, just as astronomical periods and life cycles interlock. This coordination of time scales was both objectively instantiated in the world and subjectively mirrored in the positivistic calendar, an artefact devised to remind humankind of order and progress. Comte thus came to envision "positive knowledge" as a new type of religion, one that would bring new stability and spiritual authority to a modern world in utter disarray since the French Revolution. He thus actively engaged in spreading his worldview, by writing letters and dispatching emissaries to the British Isles, Germany, Russia, and even to the Indian subcontinent and the Ottoman Empire. The closer bonds were with Victorian Britain, where the English philosopher John Stuart Mill came to appropriate a brand of positivism which he adjusted to his own domestic agenda. While endorsing the "anti-metaphysical" quest for scientific objectivity, he turned positivism into an inductive enterprise more in step with the British tradition. Mill's positivism differed from Comte's on at least two scores. While leaving aside the original emphasis on the conceptual resources of higher mathematics that were central to Comte's philosophy, Mill disagreed with his French counterpart on the very meaning of universalism. Whereas Comte explicitly and consistently condemned all forms of colonialism, slavery and purportedly Christian civilizing missions, Mill acknowledged the European civilizing burden and supported British rule over India.

Whereas Mill wove his universalism into a defense of individual liberty, extended to European women, Comte conceived of his in connection with the full

FIGURE 4.3. View of the interior of the positivist temple in Paris, the Chapel of Humanity, at 5, Rue Payenne in Paris. Wikimedia Commons.

recognition of cultural differences. More specifically, Comte did not conceive his three-stage law as the uniform unfolding of a unilineal worldwide sequence. On the contrary, universal positive science was everywhere dependent on particular social practices and cognitive capabilities that were to serve the acquisition of knowledge by adhering to the corresponding social contexts. Unlike Mill, Comte valued these cultural specificities, which he conceived as "fetishistic" worldviews. These, he suggested, might offer unexpected shortcuts to the positivist stage, while relieving large parts of humanity from the troubles associated with the metaphysical stage, which still afflicted so many Europeans. The Comtean vision of such variegated routes to universalism proved appealing to many non-Europeans who, from Brazil to British India and the Ottoman Empire, found in positivism a way to promote scientific progress without renouncing local knowledge and social forms of organization (Feichtinger, Fillafer, and Surman 2018).

BEWITCHED SYMBOLS: SYMBOLICAL ALGEBRA AND LOGIC

In May 1812, a group of Cambridge students, including Charles Babbage, John Herschel and George Peacock, came up with a novel idea, at first more in jest. They decided to translate Sylvestre-François Lacroix's *Traité élémentaire de calcul différentiel et integral*, and then to distribute it just as the Bible Society did with the Holy Scriptures. Dissatisfied with an English mathematics fettered by Newtonian orthodoxy, they founded the so-called "Analytical Society" with the aim of elevating Cambridge mathematics to French standards (Enros 1981, 1983). Their initial focus fell on notation, so they proposed to replace domestic fluxions with continental differentials, or, citing Babbage's well-known witticism, to exchange "the Dot-age of university" with "the principles of pure D-ism." In England, derivatives were indeed commonly denoted after Newton by raised dots, *viz.*,

$$y = t^2, \dot{y} = 2t, \ddot{y} = 2,$$

while on the Continent, Leibnizian ratios of differentials prevailed, *viz.*,

$$z = t^2, \frac{dz}{dt} = 2t, \frac{d^2z}{dt^2} = 2.$$

Leibniz's *d*'s were deemed more flexible than Newton's dots, for example, in the case of multiple differentiation. Leibnizian quantities were also more general, since they did not rely on taking t (time) as the sole independent variable. Moreover, the Continental calculus was a well-developed tool; its operations stood apart from the quantities it operated upon, which seemed to open inexhaustible conceptual possibilities. Descartes' innovative exponential notation, *viz.* a^2, a^3, . . ., for instance, proved to be highly compatible with the symbolism employed in the Leibnizian calculus. A whole generation of British mathematicians launched into pondering how and why notation proved crucial to mathematical thinking.

In spite of marked differences of opinion regarding the essence of algebra and symbolic mathematics, a common unitary view of mathematical truth prevailed among British mathematicians. This endowed mathematics with the same absolute certainty and necessity as the transcendent truths of religion, which they deemed to be rooted in the depths of the human heart. Babbage, Peacock and Herschel thus supplemented their translation of Lacroix with footnotes, in which they expressed their disagreement with his concept of limit as a standardized, conventional, and impersonal idea. Instead of Cauchy's rigor, they valued Lagrange's algebraic foundation of analysis as more congenial to the culturally prevalent norm of necessary truths grounded in personal knowledge, an Austenian harmonized balance of reason and feeling.

Due to its seductive fruitfulness, symbolical reasoning profoundly challenged the British conception of mathematical truth, which led to disagreements over the very essence of algebra. Refusing to acknowledge the geometrical interpretation of imaginary numbers as "an essential or necessary property of these signs," Peacock affirmed in the early 1830s that the rules for operating with the symbol $\sqrt{-1}$ sufficed to deliver all the meaning one could attach to these quantities. In response, the Irish mathematician William Rowan Hamilton demarcated three different ways to envisage algebra, either as a "practical," "philological," or "theoretical" endeavor. He thus distinguished between algebra as an art or a technique, that is a mere system of rules to be valued only for its practical usefulness; or as a language or notation, developed for its conceptual expressiveness and logical properties; or lastly, as a science proper, that is a system of truths grounded in the contemplation of ideas. In stark contrast to Peacock's approach—which he presented as reducing algebra to a mere game of symbols, "an affair of pothooks and hangers, of black strokes upon white papers, to be made according to a fixed but arbitrary set of rules"—Hamilton defined algebra as the science of pure time, grounding algebraic truths in intuition (Bloor 1981). Augustus De Morgan thus expressed a common feeling when he depicted Peacock's algebra as "symbols bewitched, running about the world in search of meaning." Yet, Babbage would venture even further than Peacock in committing himself to severing algebra from its various applications and turning it into a pure unfettered language of signs.

These interpretations of negative or imaginary numbers by means of algebraic symbols raised the problem of how to cut these loose from ordinary arithmetic when algebra derived its validity from the latter. Peacock proposed a straightforward answer to this quandary by dropping the ties to arithmetic altogether. He, thus, no longer conceived of symbolical algebra as "arithmetical algebra" or "universal algebra," in which letters were substituted for numbers to obtain generalized patterns of arithmetical truths. Even if algebraic identities like $a^2 - b^2 = (a - b)(a + b)$ were originally derived from arithmetical truths, they need not be grounded by these. Peacock's so-called "principle of permanence of equivalent forms" thus vouched for the truth of these forms, now uprooted from their arithmetical soil (Lambert 2013). Not everyone agreed, though. Acknowledging that it was not only possible but also fruitful to develop uninterpreted symbolic systems in a purely abstract way, De Morgan nevertheless insisted that formal consistency was an illegitimate goal for mathematical knowledge. His recurrent metaphor was that of a jigsaw puzzle played upside down with the promise of a picture revealed in the end (Richards 1988). In De Morgan's view, an algebraic process remained incomplete unless one carried out an appropriate interpretation; to gain acceptance, algebraic knowledge would eventually need to become no less descriptive than that of geometry.

Babbage's views were far less traditional. He saw mechanizing intelligence and rationalizing the manufactory as similar endeavors, based on division of labor, mental or physical (see Chapter 6). In both cases, a new transparency gained from a privileged standpoint led to increased efficiency. Inspiration for this came from Jeremy Bentham, an admirer of the French Enlightenment. Bentham devised a so-called "panopticon" that could be introduced anywhere, whether in the barracks or in the factories, yielding "the diagram of a mechanism of power reduced to its ideal form" (Foucault [1975] 2012). In response to Whewell's contention that analytical mathematics did not belong to liberal education, since only induction could procure insight into the divine plan of the world, Babbage replied with an unorthodox theory of miracles. One of his favorite tricks when presenting his calculating engine consisted in adjusting it so that it would spit out the first million integers in order, then start counting by ten thousands before some sequence like the first continues again. Whereas these discontinuities appeared as unexplained "miracles" to the spectators, they were actually programmed into the machine from the start. Babbage drew the far-reaching conclusion that miracles were not a breach of established laws, but rather they testified to the existence of higher-order laws that produced pre-intended results at an appointed time (Schaffer 1994). In a notebook of 1837, Charles Darwin recorded his own faith in a miracle-free universe. Considering whether "creative power acted at Galapagos," he noted "this might be made very strong, if we believe the Creator creates by any laws, which, I think is shown by the very facts of the Zoological character of these islands."[4] A later entry, dated December 2, 1838, bears witness that he read Babbage's Treatise in favor of a lawlike universe. Spotting the passage where Babbage quoted from Herschel, Darwin noted in this connection: "Herschel calls the appearance of new species the mystery of mysteries, & has grand passage upon the problem!—Hurrah. 'intermediate causes'."[5]

George Boole proposed another way to solve the conceptual tension inherent in the whole enterprise of symbolical algebra, though his ideas met with ambivalence. "They who are acquainted with the present state of the theory of Symbolical Algebra, [Boole observed,] are aware, that the validity of the processes of analysis does not depend upon the interpretation of the symbols which are employed, but solely upon the laws of their combination."[6] Hence, these laws should be detached from their arithmetical origin and founded upon the mental processes they represent. In his *Mathematical Analysis of Logic* (1847), Boole contrasted two views of logic, *viz.* "from without," with reference to quantity, number and magnitude, and "from within," meaning based upon facts grounded in the constitution of the mind. Understood as the mathematics of the human intellect, his algebra of logic aimed at establishing the "speculative truth" of such laws originating in "the existence in our minds of general notions, our ability to conceive a class, and to designate its individual members by a common name."

In that same year of 1847, however, De Morgan explicitly acknowledged the inner limitation of the logical theory of syllogisms. Arguments of the type "man is animal, therefore the head of a man is the head of an animal" were indeed inferences, but not syllogisms—a logical fact that would later be accounted for by a different logic based on quantifiers. In his *Investigation of the Laws of Thought* (1854), Boole went a step further. Symbolic logic, conceived as the science of mental operations, would also form the basis of a theory of probability, which therefore belonged equally to the science of number and logic. The rationale for granting logic priority over probabilities lay in that we can express the sought probability of an event as a logical function of the events whose probabilities are given. Boole's logic was thus essentially two-sided, and its symbolic products and sums capable of interpretation in terms of classes or in terms of probabilities.

NON-EUCLIDEAN GEOMETRIES AND PHYSICAL SPACE

Throughout most of the century, natural philosophers and mathematicians implicitly admitted that physical space came endowed with a Euclidean geometry. Light rays and force-free motions provided empirical evidence supporting this view, since both served as physical instances of geometrical straight lines. On the theoretical side, Kant's robust epistemology had granted Euclidean truths the status of synthetic a priori propositions (see Introduction). This Kantian legacy gradually ceased to remain unquestioned; nevertheless, it enduringly shaped philosophical debates on science (Friedman and Nordmann 2006). In Britain, Euclidean geometry long remained a particularly unassailable stronghold. The view of Euclidean space as an external reality had gained theological support, on account of Newton's conception of space as co-extensive with a ubiquitous God, as well as the cultural pervasiveness of natural theology in Victorian scientific circles (Richards 1988). Besides, Euclid's

Elements dominated the British curriculum until at least the mid-1860s as the unrivalled manual for teaching geometry, a central topic in a liberal education. Euclid's critics were very few at the time. At the 1869 meeting of the British Association for the Advancement of Science, James Joseph Sylvester jibed at those "who ranked Euclid second in sacredness to the Bible alone." In later years, an association for the improvement of the teaching of geometry formed, led by Thomas Archer Hirst. Nevertheless, all attempts at challenging Euclid's prevalence failed, in part because of the adamant resistance of such towering figures as Arthur Cayley (Moktefi 2011).

For reasons related to their entrenched mistrust of merely analytical forms lacking geometric interpretation, the British long ignored Riemann's infinitesimal approach to general geometries (see Chapter 3). Things changed when Helmholtz took up Riemann's ideas in the late 1860s and made them digestible for a British audience. As a physiologist, Hermann von Helmholtz was interested in identifying the experiential basis of our knowledge of space. In Germany, at that time, two groups were engaged in a hot debate over theories of spatial perception. The so-called nativists held that spatial notions were inherent in our very sensations and developed through physiological mechanisms within our minds. The empiricists, led by Helmholtz, advocated an opposing view, which considered sensations to be merely arbitrary signals from which the mind produced a representation of space through perceptual experience. With this background, Helmholtz envisioned a connection between Riemann's mathematical approach and his own empirical position. His contention was that in early childhood one acquired basic spatial knowledge through experience by moving rigid bodies around in space. Helmholtz then argued that this very possibility of having a space that allows arbitrary motions of rigid bodies without deformation actually required that the space have constant curvature. This hypothesis effectively cut back the whole range of possible Riemannian worlds to those three-dimensional geometries with a fixed curvature, the so-called classical non-Euclidean geometries. In this setting, however, the axioms were no longer necessary truths but rather the condensed expression of a fundamental fact of experience. Thus, only actual measurements could suggest which geometry favored our experience.

At about the same time, a violent controversy broke out in Imperial Germany, as new ideas from Britain spread across the Channel. In 1872, Leipzig professor and astrophysicist Carl Friedrich Zöllner launched an assault against the Berlin scientific elite, alleging they had sold out to the British natural scientists. He aimed his fiercest charges at Helmholtz, who supervised the German translation of Thomson and Tait's *Treatise on Natural Philosophy*, in Zöllner's eyes a paragon of appalling British inductivism and a strong dissolvent of his own grand worldview heavily imbued with *Naturphilosophie*. Whereas Helmholtz supported Maxwellian field physics, Zöllner advocated Weberian action-at-a-distance electrodynamics. Here a cosmopolitan view of science clashed with an approach strongly coupled with nationalist yearnings. As an outspoken censor of outdated Romanticism, Helmholtz praised the British for having purified natural science from abstruse metaphysics. He dismissed the "Icarus' flight of imagination" in favor of honest mathematical toil.

Zöllner's ideas, however, elicited responses beyond scientific circles. His model of a finite, unbounded, space of constant positive curvature caught the attention of young Friedrich Nietzsche, whose notebooks touched upon Olbers' paradox and its purported cosmological solution. Zöllner later turned to Spiritism with the hope of proving the existence of a fourth dimension experimentally. He leapt to this speculation based on Felix Klein's offhand observation, which he took literally, that simple closed curves can always be unknotted in a four-dimensional space (Epple 1999: 165–66). When the American medium Henry Slade came to Leipzig in 1877, Zöllner organized Spiritistic séances, which he purportedly controlled. A fierce public debate ensued, hinging on the criteria required for advancing an experimental proof, *viz.* involving issues such as human perception compared to the sensitivity of measuring instruments as well as the authority of the experimenter. After Leipzig, Slade continued his European tour in Saint Petersburg, where his performances prompted inflamed exchanges in the Russian press. While the chemist and Academician Butlerov claimed that modern mathematics could account for a fourth-dimensional reality causally impinging on our material world, another chemist and father of the periodic table, Dimitri Mendeleev, formed a committee to reject such claims. Feodor Dostoevsky also entered this arena when he made his character Ivan Karamazov recall that God created a Euclidean world and provided us with "a Euclidean earthly mind." Zöllner published an open letter in support of Butlerov. Drawing on Riemann's recently published fragments on *Naturphilosophie*, he argued that atoms of matter would be the entryways to the fourth dimension in a three-dimensional world.

This episode reveals a fault line separating the proponents of two conflicting views of physical space in relation to geometry. Adding to the confusion, some on both sides failed to recognize that Riemannian curvature was an intrinsic property. They then imagined a curved three-dimensional space embedded "within" a four-dimensional reality to which it was causally connected, whereas others thought of these "spaces" as mere analytical constructions. Inspired by the physical ideas underlying Riemann's approach, William Kingdon Clifford escaped this dilemma. Endowing spaces of variable curvature with a physical meaning, he made them the spearhead of his attack against necessary law and the cornerstone of his own agnostic worldview.

LIMITS OF KNOWLEDGE, ENTANGLED WORLDVIEWS

On August 14, 1872, the renowned Berlin physiologist Emil du Bois-Reymond gave a plenary lecture on the limits of natural science at the annual conference of German scientists and physicians held in Leipzig. One year after the proclamation of the new German Reich in Versailles, his advocacy of restraint had a political ring to it. Natural science, "the world-conqueror of our times," had to acknowledge the true boundaries of its empire. In accordance with the mechanistic worldview of his youth, a view he shared with Helmholtz and the whole generation of the so-called "organic physicists" of the late 1840s, du Bois-Reymond claimed that the resolution of natural processes into the mechanics of atoms now set the very standard for any scientific explanation. Formulated by means of mathematical equations, he granted the propositions of mechanics the same apodictic certainty as those of pure mathematics. This ideal of knowledge thus amounted to the framing of a system of differential equations from which, in principle, one could determine at any given time the position, velocity, and direction of all the atoms in the universe with complete accuracy.

There were obstacles, however, that even a Laplacian demon could not overcome. Du Bois-Reymond identified two main difficulties marking the limits of natural knowledge, *viz.* the nature of matter and the relationship between brain states and consciousness. On both questions, natural science fell short of providing a comprehensive scientific worldview. It thus failed to account for its own presuppositions, as well as to yield a unified worldview of matter and consciousness. As regards these enigmas, the natural scientist ought therefore not only to avow that we do not know, but furthermore that we will not know.

Whatever du Bois-Reymond's ultimate motivations were for urging his contemporaries to renounce extravagant claims to naturalistic omniscience, many regarded his lecture as a political statement sounded in the midst of Bismarck's *Kulturkampf*. Its repercussions within a recently united Germany suggest the degree to which scientific and religious worldviews reflected ambivalent attitudes regarding Bismarck's aggressive efforts to create a Protestant state. A heated controversy indeed ensued with opponents ranging over a whole spectrum of positions, from those who disclaimed the Laplacian ideal to those who embraced it, but refused its purported limitations. The very last word—"*Ignorabimus!*"—with which du Bois-Reymond concluded his address became "a kind of *naturphilosophisches* Schibboleth," and the whole battle between "Natural science versus *Naturphilosophie*" came to be known as the *Ignorabimus-Streit*.

A decade later, the mathematician Paul du Bois-Reymond, younger brother of the physiologist, brought forth a different kind of unsolvable enigma inherent in the mathematical concept of limits. These arise in connection with infinite sequences, thus matters rooted in the very nature of representations involving the thought-processes of pure mathematics. Since Cauchy, mathematicians had become accustomed to considering infinite sums that converge toward definite values under certain conditions, as for instance $\frac{1}{2}+\frac{1}{4}+\frac{1}{8}+\cdots$. However, du Bois-Reymond distinguished between two mutually exclusive views regarding what it means for a limit to exist. Advocates of "idealism" believed in the reality of the limiting concepts required by our thought-process, even though these elude either perception or representation. Those who adopted "empiricism," on the other hand, acquiesced to a "system of complete renunciation" by taking the view that to claim something exists requires that it somehow be perceptible. Unlike the "empiricist," the "idealist" feels confident in passing to the limit provided the required conditions hold, even though there is no way to have direct access to that limit. Exactness in measurement makes perfect sense to "idealists," but not to "empiricists." The idealist's notion of a continuum, whether arithmetical or geometrical, composed of discrete points, appears utterly absurd to the empiricist. These irreconcilable views, each laying justified claims to qualify as the fundamental intuition underlying the exact sciences, led to a "dualism of the worldview,"[7] a no less enthralling phenomenon, du Bois-Reymond claimed, than the double image seen through a birefringent crystal. Charles Hermite acknowledged that the argument was puzzling:

> You hold me in the lurch, [*he wrote to du Bois-Reymond,*] I am suffering Tantalus's torment, with this mysterious expression in which until now I saw no harm, of the infinite decimal development : 0, $\alpha_1\ \alpha_2\ \alpha_3 \ldots$ Shifting from arithmetic to algebra, I wonder whether in any convergent series, there would be similarly "*in cauda venenum.*" In short, thanks to you, I see the entire mathematical edifice falter on its foundations threatened with impending ruin, and like M. Renan who says he does not know whether he is a materialist or a spiritualist, as for me, I doubt whether I have to declare myself an idealist or an empiricist.[8]

These were difficulties of a philosophical, rather than a mathematical, kind. In practice, one could always adopt a neutral approach by way of eclecticism, following du Bois-Reymond's catchphrase: "Empiricist language, idealist proofs." Yet, the enigma would remain unsolved and the core antagonism between "idealism" and "empiricism" would even extend from pure mathematics to natural science. While first entangled in the dual worldview presumably inherent in mathematics, these modes of representation led to openly conflicting worldviews when shifted to natural science. In du Bois-Reymond's view, "we are enclosed in the shell of our perceptions and born blind for what is beyond them."[9] The reality of the world is definitely beyond our reach and, owing to its dual character; mathematics cannot help us break this shell.

DETERMINISM AND FREE WILL

At about the same time, in France and Belgium, advanced mathematics unexpectedly came to occupy center stage in philosophical debates on reductionism, determinism and moral freedom. These took place against the backdrop of increasing social, religious and political tensions. A major shift had occurred in the late 1860s, when the physiologist Claude Bernard introduced the modern concept of "scientific determinism." In contrast to the grand metaphysical view of causality resting on Leibniz's principle of sufficient reason, which underlay the Laplacian demon metaphor, the new concept dispensed with such ontological assumptions. It merely claimed that "initial conditions" bring about, hence "determine," specific phenomena. Although Laplace himself never used the word, so-called "Laplacian determinism" had a different meaning. Rooted in the belief that nature is ultimately knowable, Laplace developed a heuristics based on the calculus of probabilities as the indispensable tool for disentangling regular from accidental causes and disclosing the regularities of nature. By the end of the century, however, contemporaries had come to understand Laplace's metaphor as a full-fledged materialistic determinism (Cassirer 1956).

As the anticlerical French Third Republic solidified, Catholic scholars, still influential within academia, were anxious to resist the seemingly inexorable demise of Christian faith. Concerned with reconciling man's moral freedom with nature's mechanistic laws, they sought to assuage the conflict between science and religious beliefs. Among them, Joseph Boussinesq connected the philosophical issue of determinism with an apparent paradox, first noticed by Poisson in 1806, namely the existence of multiple solutions at singularities for certain kinds of differential equations. Boussinesq saw these so-called "singular solutions" as offering a unified descriptive framework accounting for organic life and free will, while meshing with the deterministic mechanical processes of inanimate matter, but without reducing the former to the latter. By contrast, "Laplacian determinism" would appear with hindsight as based upon the existence and unicity of solutions in the case of ordinary differential equations akin to Newton's law, a theorem proved by Cauchy in the 1830s. Unicity made it possible to assert knowledge of the evolution of a mechanical system in the future as well as the past based on knowing the present positions and velocities of each particle. In the general case, however, the existence of singular solutions would imply bifurcating paths at branch points of sorts, making indeterminism appear internalized within mathematics itself. This, Boussinesq suggested, left room for the intervention of an extra-physical "directing power" orienting mechanical action but without disturbing the laws of matter. Arguing against presupposing vital forces in addition to physico-chemical ones, Claude Bernard and Augustin Cournot had previously compared the mechanical action of the will to a negligible quantum of work, like slowly pressing the trigger of a gun. Boussinesq, however, innovated in observing that mathematics made even such infinitesimal "trigger action" superfluous, since free will could act seamlessly on matter without encroaching upon it. His suggestion prompted responses from

Adhémar Barré de Saint-Venant, his mentor, who reiterated Cournot's original argument, and Joseph Bertrand, who scathingly censured an allegedly misguided conception of the relationship between mathematics and the real world

In Victorian Britain, James Clerk Maxwell also honed a set of arguments against the rising materialist mechanistic worldview by developing a similar metaphor. In 1862, he observed that body and soul do interact, though without exchanging energy, rather "as . . . when a pointsman shunts a train, it is the rail that bears the thrust."[10] Borrowing from Balfour Stewart, he conceived stable physical systems as deterministic, whereas unstable ones, like balanced eggs tipping over, allowed "freedom of action." The same metaphor would later be expanded to "pick a hole" in the second law of thermodynamics, with the steering power of the pointsman conferred to Maxwell's demon.[11] Maxwell acknowledged that Boussinesq "had managed to reduce [*all trigger-work*] to mathematical zero," so that "when the bifurcation of path occurs, the system, ipso facto, invokes some determining principle which is extra physical (but not extra natural) to determine which of the two paths it is to follow."[12] Although sharing common scientific practice to a considerable extent, two distinct Victorian worldviews emerged that sharply differed on how to circumscribe such naturalness. The theistic position, championed by Maxwell, here stood in opposition to the natural scientific views of Thomas Henry Huxley, nicknamed "Darwin's bulldog" (Stanley 2015; Harman 1998).

Across the Channel, other dividing lines shaped the intellectual landscape. Whereas the Comtians had striven to eliminate metaphysics from science, Catholic circles led a powerful anti-positivistic reaction in both science and philosophy, typified by Jules and Paul Tannery and Pierre Duhem. In this respect, Émile Boutroux's thesis *On the Contingency of Natural Laws* (1874) represented an important turning point. Many in this generation had opposed scientism, but as reported by the novelist Paul Bourget, an acute observer of the French intellectual arena, Boutroux was the only one who spotted its Achilles heel (Nye 1979). Laws are artificial creations of the mind. They are none other than contingent abstract mathematical frameworks failing to capture any superior reality or rule things. "Laws are the channel along which rushes the stream of facts: these latter have hollowed it out, although they follow its track."[13] There is no analogue to mechanical laws in reality and therefore, according to Boutroux, Boussinesq's argument missed the point since "there is no chain to be broken in order to enable a moral influence to permeate what is called the world of matter and motion."[14] In contrast to Boutroux, his brother in law, Henri Poincaré, disclaimed the view that laws were ontologically contingent as an exaggerated nominalism. In the long run, he argued, scientific laws would help us grasp some structural features of objective reality. Yet, to a certain extent, his moderate conventionalism was akin to Boutroux's thesis in a secularized form. Another aspect of Boutroux's philosophy was influential, as he also contributed to initiate French historical epistemology. Since experiment is the only source of science, knowing things historically was in his view the only way to know them objectively (Rheinberger 2010).

MATHEMATICAL MODERNISM AND *FIN-DE-SIÈCLE* ANXIETIES

In a book published in 1890, one year after his untimely death, the young Strassburg philosopher Benno Kerry clearly identified the source of a fundamental disagreement regarding the ontology of mathematics. Based on a thorough analysis of the thought processes involved in the concepts of limit, he challenged du Bois-Reymond's presumably ultimate "dualism of the worldview." The question whether the limit of an infinite decimal does or does not exist, he claimed, should have a clear-cut answer, irrespective of one's "idealist" or "empiricist" leanings. Unsatisfied with what du Bois-Reymond allegedly celebrated as an "inescapable and impregnable dilemma," Kerry refused to admit that this would-be antinomy lodged in mathematics was tantamount to the successful epistemological solution itself. He was among the first to diagnose clearly that the heart of the matter lay in the strength of language. The mathematical existence of objects thus required having definitions based on non-contradictory concepts subject to consistent application. Language, however, involves more than meets the eye, and sometimes contradictions subsist in our concepts, in very subtle ways. "Language is a sword as much guided by error and fraud, as by truth," Kerry noted, and "so long as the ideal of a universal conceptual language is not achieved, one can only find shelter from its snares by identifying them with warning lights."[15]

At the close of the nineteenth century, this kind of recasting of mathematics into an autonomous self-defining discourse had become pervasive, and this led to an epistemic breach that separated those who resisted from those who welcomed it. Herbert Mehrtens retrospectively made this the very cornerstone of his interpretation of mathematical modernism. "The new language of mathematics does not need to be ascertained by an external reality, because it ascertains itself through its own constant work" (Mehrtens 1990: 68). Many major actors of the period would, on this view, fall into two distinct cohorts. Whereas the "moderns," championed by David Hilbert, proclaimed logical consistency was the bedrock of mathematical truth, the "counter-moderns," including Felix Klein, Henri Poincaré and L. E. J. Brouwer, sought to anchor mathematical knowledge in intuition and the pre-linguistic activity of the mind. In spite of its being too starkly unequivocal, this interpretative grid proved to be powerfully suggestive, paving the way for later works, such as Jeremy Gray's study (Gray 2008), which adopted a more encompassing view of the "modernist" transformation of mathematics.

Modern mathematics chiefly implied a turning point in the ontology of mathematical objects (Gray 1992), which Ernst Cassirer later analyzed in depth:

> The existence of number *e* means nothing else than that, within the ideal number system, one and only one position is determined definitely and with objective necessity by the series, which we apply in its definition. If we assume the general rule, $1+\frac{1}{1}+\frac{1}{1.2}+\frac{1}{1.2.3}+\ldots$, then by it the system of rational numbers is analysed into two strictly divided classes, of which one contains all the elements, that are ever exceeded by the series when it is carried far enough, while the other contains all those elements, with which this is not the case. (. . .) The validity of all these relations alone justifies us in speaking of a "number" *e*, and constitutes the entire "being," the complete and self-contained existence of this number. (Cassirer [1910] 2003: 124)

While intuition gradually fell from grace as a reliable basis for mathematical knowledge, logic came to the fore as the main driving force of modern mathematics. From Weierstrass' example of a continuous, but nowhere differentiable function, to Peano's and Hilbert's space-filling curves, examples abounded of mathematical objects that eluded all grasp by means of intuition (Lützen 2003). This ontological and epistemological shift was a characteristic feature of mathematical modernism, but many hardly saw this outcome as a sign of progress. Those who embraced it saw the gains made in rigor that came along with autonomous abstraction, but also a certain sense of anxiety (Gray 2008). Even Cantor expressed disorientation in an early letter to Dedekind, written in June 1877. He urged him to look at his newly found proof of a one-to-one correlation of manifolds of different dimensions, wondering whether it was really "arithmetically rigorous." Until he received confirmation, he wrote

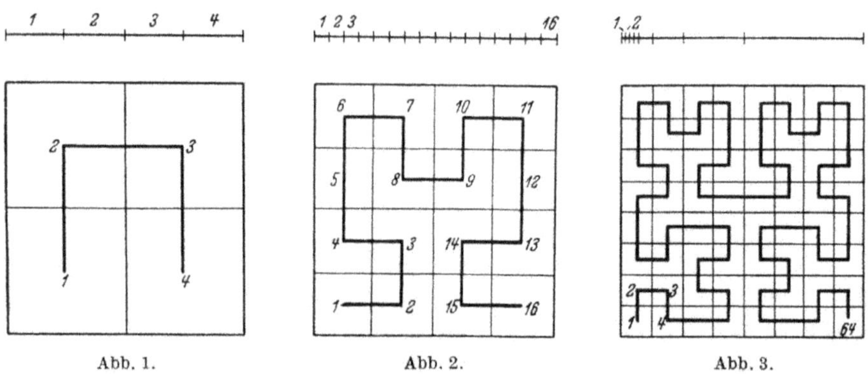

FIGURE 4.4. Hilbert's procedure for constructing a space-filling curve, *Mathematische Annalen* 38 (1891), 459–60. Image courtesy of David E. Rowe.

that he could only say, "I see it, but I do not believe it" (Ferreirós [1999] 2007). Others expressed even stronger dismay. Hermite wrote to Stieltjes that he turned away "with fear and horror from the lamentable plague of those functions which do not have a derivative," while Poincaré ironically observed that "logic sometimes breeds monsters."

In his 1879 *Begriffsschrift*, Gottlob Frege developed a conceptual notation that enabled him to express the logical structure of mathematical concepts and inferences with the goal of reducing mathematical knowledge to pure logic. While thus taking up Leibniz's dream of a *lingua characteristica* and a *calculus ratiocinator*, Frege's main goal was to exhibit the conceptual fruitfulness of the logic of relations. Unlike Boole's logic, in which propositions remained unanalyzed

FIGURE 4.5. Felix Hausdorff (1868–1942). By permission of the Hausdorff Edition, Bonn University.

and only expressed relations in terms of truth values, Frege's logic could articulate meaning and represent the inner structure of propositions by means of predicate letters, variables, and embedded quantifiers. In Frege's view, Boole's logic offered a sentential calculus, but failed to deliver a quantification theory, whereas Peano's logic had a notation for quantifiers, but poor inferential derivations. Only his own system would present both aspects built into the same formalism.

By the turn of the century, the mathematician Felix Hausdorff, a paradigmatic figure in the emergence of mathematical modernism later known for his axiomatization of topological spaces, gradually took up Cantorian set theory. This interest arose, however, while he was pursuing philosophical investigations in another key and under another name, as Paul Mongré, his second *persona*. In 1897, Mongré, the literary critic and Nietzschean philosopher, wrote a collection of aphorisms in which he developed his epistemological views. A recurrent theme was the human inclination to ascribe as features of the objective world certain subjective impressions resulting from selective sensory perceptions. One of Mongré's metaphors involves an orchestra, whose instruments "rage undirected in a disharmonious mess," from which intermittent

FIGURE 4.6. Ernst Mach (1838–1916), *c.* 1910. Photography by Charles Scolik. Photo by Imagno / Getty Images.

listeners would extract "a rhythmically clear, pure and harmonious melody" (Hausdorff's *Sant'Ilario* 1897, quoted in Epple 2006: 270). Taking up this motif on a cosmic scale, Mongré argued that even a perfect cosmos could arise from the wildest chaos through the selective sieve of consciousness; this being so, nothing could be truly known about the world itself. Nietzsche's doctrine of eternal recurrence captured his interest, and he tried to refute the tentative proofs found in Nietzsche's *Nachlass*. Three notebook entries from the early 1880s, in which the philosopher discussed the notion of a spherical space, the possibility of a recurrent state of the world, and Olbers' paradox, indeed attest to the fact that he carefully read Zöllner (Small 2017).

In refuting Nietzsche's purported proof, Mongré soon realized that he could find useful ammunition in Cantor's point set theory. Within a year, he embraced this new conceptual arsenal as a tool for a self-critique of science. In 1903, the mathematician Felix Hausdorff took up the issues first raised by Mongré, using mathematics to elaborate an original theory of space founded upon the systematic exploration of the range of variations allowed for our mathematical concepts within the bounds of experience. Set theory and Hilbert-style axiomatics provided exactly what he needed for this. Modern mathematics thus unexpectedly contributed to shape *fin-de-siècle* perspectivist epistemological views.

MANY WORLDVIEWS, ONE WORLD-PICTURE

At the end of the nineteenth century, the dominant mechanical worldview had been already seriously challenged by the Austrian physicist and philosopher Ernst Mach, whom Einstein would later credit for having "shaken this dogmatic faith." In his *Analysis of Sensations* (1896), Mach referred to Emil du Bois-Reymond's *Ignorabimus* as a necessary, although incomplete, step toward emancipation from a persistent dogma. The Berlin physiologist nevertheless failed to overthrow it, for "like countless others, he took the instruments of a special science to be the actual world." In the 1880s, Mach regarded the Laplacian ideal, "the projected worldview of the Encyclopedists," as "a mechanical mythology." By contrast, he observed that "natural science does not pretend to be such a complete worldview," for "the highest philosophy of the natural scientist consists precisely in tolerating an incomplete worldview and in giving it preference over an apparently self-contained, but inadequate one."[16] In accordance with his economical conception of science as a process of gradual adjustment of our thoughts to our perceptions, Mach denied from the start that one could take a full-fledged worldview for granted, as if it were a gift. The only possibility for gradually approaching the ideal of a unified worldview was through the hard work and toil of generations of researchers. This Machian view, however, soon became a point of contention.

FIGURE 4.7. Max Planck (1858–1947), 1919. Photo by Bettmann / Getty Images.

On December 9, 1908, Max Planck gave a lecture in Leiden on this very topic of the unity of the physical world-picture,[17] but contrary to Mach, he claimed there was no such thing as an approximate picture of reality. In Planck's view, one could never reach a coherent physical world-picture, properly understood, in a stepwise or piecemeal fashion. Significantly enough, he substituted the word "world-picture" [*Weltbild*] for "worldview" [*Weltanschauung*], as previously used by Mach. In the older physical system, a particular picture prevailed for each class of phenomena, which resulted in a mosaic of mutually inconsistent pictures, though removing any one of these would compromise the others. For Planck, none of this pertained to a genuinely scientific "world-picture." In such a system, "no single trait can be left out as inessential, each is rather an indispensable constituent of the whole and possesses as such a definite significance for observed nature, and the other way around, every observed physical phenomenon will and must find its exact corresponding place." Far from being a more or less arbitrary creation of our minds, the "world-picture" ought to reflect real natural processes existing in complete independence from us. Again, in opposition to Mach, Planck vindicated the reality of atoms, which he held were "no more and no less real than celestial bodies or the earthly objects around us." He explained the success of Mach's ideas by the sense of disillusionment resulting from the excessive expectations attached to the grand mechanical worldview of the previous generation. However, in Planck's view, Mach "overshot the mark in downgrading the physical world-picture together with the mechanical one."[18] While avowing his "aversion to [what he called] hypothetico-fictive physics," Mach mockingly replied to Planck's charge of being "a false prophet" by turning his back on the "church of physics" which had proceeded against him "after the fashion of Pius X," the Pope who condemned "modernism" as a heresy in 1907 (Blackmore 1992).

Planck nevertheless acknowledged that with the introduction of probability in physics, "the hope of completely working out the mechanics of atoms had been fundamentally abandoned." Still, this could not mean, in his view, renouncing any world picture. Boltzmann, for his part, claimed that the task of theory consisted in constructing a mental picture of the world distinct from experience. This could then guide the researcher in thought and experiment, even if intricate mathematics were required in the process. Hence, Boltzmann deemed the purely descriptive ideal of mathematical phenomenology an illusion, which accounts for the emphasis he placed on the heuristic role of mathematics in the construction of these mental pictures (Blackmore 1995; de Courtenay 2018). At the end of the 1880s, in a series of articles published in the *Philosophical Journal*, Boltzmann stood accused by P. G. Tait of turning the physical sciences into an empty play with "terrific arrays of symbols" (Bellone 1980).

Orthodox British methodology had long viewed mathematical representations in physics as no more than idealizations of common sense; they could only convey knowledge obtained inductively. Indeed, throughout the second half of the century, the British strongly subscribed to the explanatory power of mechanisms with which they devised a wealth of refined models. In contrast to British physics, which Pierre Duhem suggestively depicted as all pulleys, strings, wheels, tubes and hooks,[19] Boltzmann explored pathways by which mathematics might help in constructing better images. In a short tract titled "More on Atomism" (1897), he showed how, on closer inspection of the very meaning of mathematics, a different interpretation of Fourier's heat equation was possible. Let us imagine, he proposed, a large number of finite points, like the vertices of a set of regularly stacked cubes. Suppose further that an initial given temperature is associated with each point. With the passage of time, the temperature at each point will approximate the arithmetic mean of the temperatures at the six neighboring points. The same then applies to any of the six neighboring points, and so on. This sequence of computational instructions, Boltzmann argued, yielded a clear conceptual picture that was unobjectionable, atomistic, and epistemologically superior to any other. Should we cling to the old phenomenological view and imagine a body consisting of elements of volume with the temperature at each point being a continuous function of the coordinates that satisfies a certain partial differential equation? By doing so, we would be unable to extract any applicable rule from the mathematical description of the situation, unlike the previous picture, which can explain the meaning of the equation by means of a sequence of instructions dealing with a finite number of elements. Boltzmann gave a very simple analogy to support this viewpoint:

> if I tell somebody to sum the series $1+\frac{1}{2}+\frac{1}{4}+\frac{1}{8}+\cdots$ to the full extent of infinitely many terms, he will be unable to do it; but if I tell him to sum so many terms that a further increase will no longer noticeably influence the result, I have given him a clear and executable prescription. All proofs that the sum of infinitely many terms equals *2* merely signify that if you add countless thousands of further terms you will never exceed *2*, though you will approach it more and more. [. . .] The same, it seems to me, holds of [. . .] the most complicated equations of mathematical physics.[20]

What mathematical equations ultimately mean may thus not necessarily be what meets the eye. With a different perspective, Henri Poincaré also tackled this same problem.[21] He observed that although new theories constantly eclipse previous ones—just as Maxwell's theory came to be preferred over Fresnel's—the differential equations that arise would remain true. This was the case because they express relations, which still reflect an underlying reality, whatever concrete images we associate with them due to changing theories over the course of time.

By the early twentieth century, *fin-de-siècle* anxieties had gradually given way to disillusionment and hostility to rational knowledge, both inside and outside academia. Many shared the conviction that the exact sciences had failed to provide an all-encompassing worldview, as theology and philosophy had done for previous generations. In this context of cultural pessimism, Robert Musil struck a different note. Mathematicians would offer the antidote to such disheartening drumbeating for feeling against intellect. "After the Enlightenment the rest of us lost our courage," Musil deplored; "a minor failure was enough to turn us away from reason," and now "we allow every barren enthusiast to inveigh against the intentions of a d'Alembert or a Diderot as mere rationalism" (see Chapter 7 of Volume 6).[22]

Ernst Mach, to whom Musil had devoted his only academic work,[23] had avowed "a lifelong melancholic nostalgia for the eighteenth century."[24] An admirer of Voltaire, he claimed the Enlightenment Age his spiritual home. Although disowning Laplace's mechanistic worldview, he nevertheless praised his demon metaphor as "a jubilant toast" to a scientific ideal "to which we too owe our intellectual freedom."[25] A member of the Vienna Circle, Philipp Frank stressed Mach's commitment to Enlightenment ideals and portrayed him as an Enlightenment philosopher for the current age. Advancing a critical attitude aimed at denouncing the misuse of auxiliary concepts, Mach's view of

Enlightenment involved a historical dialectic. Enlightening insights, gained against strong resistance at one stage, can turn into fossilized thoughts by the next (Uebel 2021). A sharp critic of the materialistic worldview of the eighteenth century, Mach thus remained faithful to Enlightenment ideals in refuting "that part of Enlightenment which was not Enlightenment."

Science, according to him, was essentially a matter of "mental economy," which mathematics honed to perfection. In pursuing a path to true Enlightenment, historical sense helps dispel the magical aura allegedly hovering over science. Elaborating on Euler's witticism that in mathematics our pen is often smarter than we could ever be, Mach observed, with a marked nod to Babbage, that mathematical signs materially disburden the mind by transferring the necessary operations to the hand. Through unpacking what scientists obliviously condense in their mathematical symbols, historical Enlightenment held the promise of an invigorated rationalism. In the next generation, as the perils of anti-rationalism loomed large, Rudolf Carnap and the Vienna Circle would find inspiration in Mach's teachings for their own scientific worldview (Carus 2007; see Chapter 4 in Volume 6).

CHAPTER FIVE

Describing and Understanding the World

From Probability and Statistics to Field Theory

SCOTT A. WALTER

The final quarter of the eighteenth century saw its share of revolutions. The North American colonies of the English Crown rebelled, and with the timely assistance of French troops and naval forces, they gained their independence, giving way to the United States of America. A few years later, across the Atlantic, King Louis XVI of France was guillotined, making way for the first French Republic. Among the changes brought about by the new French state was a closer alignment of science and technology with state interests, including the establishment of a variety of standards enforced by the power of the state. Perhaps the most celebrated of the new French standards was the metric system of decimalized weights and measures. With the assistance of members of the Paris Academy of Sciences, the meter, gram, and liter were prescribed for the measurement of length, mass, and liquid volume.

While French leaders had hoped that Britain and the United States would adopt the metric system, their proposal was flatly refused. Nonetheless, employment of the metric system became a legal obligation in France in 1795, setting in motion the standardization of units of measurement in science and commerce that expanded both in purview and in practice over the next two centuries. Scientists at the Paris Academy of Sciences came up with the first metric system, and in the nineteenth century, German and British scientists would help extend the system, recognizing its utility for communicating scientific results.

The early years of the French Republic produced important institutional changes in the organization of science and technology, including the foundation of the École polytechnique, staffed with some of France's most illustrious *savants*, and the Bureau of Longitudes, likewise staffed, and charged with publishing ephemerides for use in maritime navigation. In 1800, the French Ministry of the Interior created a Bureau of Statistics charged with quantifying selected items in topography, population, social situation, agriculture, industry and commerce. The Republic was soon overthrown by Napoleon, who had great plans for both the French state, and for all of Europe, involving military conquest, but also a thorough reform of the legal system: the Napoleonic Code. In this latter respect, Napoleon saw himself as realizing one of the primary objectives of the French Revolution: to enact a single code of law for the entire nation, a legal reform implemented by a centralized government bureaucracy.

These innovations in standardization of measurement and bureaucratic centralization, realized during the first two decades of the post-revolutionary French state, aimed to bring about greater prosperity and well-being to the populace at large. But how were prosperity and well-being themselves to be measured? An Enlightenment thinker, Jean-Jacques Rousseau, had suggested in his influential book, *On the Social Contract* (1762), that counting heads is a metric for good government:

> The government under which . . . the citizens do most increase and multiply, is infallibly the best. Similarly, the government under which a people diminishes in number and wastes away is the worst. Experts in calculation! I leave it to you to count, to measure, to compare. (Porter 1986: 2)

One might take issue with the criterion proposed here by Rousseau, and indeed, in 1798, Thomas Malthus suggested that an unchecked increase in human population size would lead to greater misery. There was, nonetheless, at the turn of the nineteenth century a strong belief in the capacity of the government to direct its resources more effectively based on precise statistical data, which led it to expend more and more resources on the collection of such data.

If the origins of what the historian T. M. Porter called the "great statistical enthusiasm" of the early nineteenth century (Porter 1986: 27) arose in France, the gathering of statistical data was also gaining momentum in Britain. Part of the motivation for this movement came from the Napoleonic Wars, as accurate census data were required in order to obtain satisfactory results from orders of conscription designed to expand the armed forces. Another factor behind British data-collection efforts stemmed from social dislocation brought about by industrialization, and a concomitant fear of a French-style revolution on British soil. In Britain, as in France, the "experts in calculation" took an interest in the numbers produced by the new data-collection schemes. Led by Malthus,

Charles Babbage, and others, the British Association for the Advancement of Science (BAAS) formed a statistical section in 1833.

INTERPRETING STATISTICS AND CALCULATING PROBABILITY

It is one thing, of course, to collect statistical data, but quite another to extract useful knowledge from that data once collected (Stigler 1986). While the problem is a very general one, it arose in a clear manner not in the domain of military conscription or social welfare, but in astronomy, which was undergoing a revolution of sorts after William Herschel's discovery, in 1781, of a previously unseen planet in the Solar System: Uranus. Using a homemade reflecting telescope, Herschel, a German-born orchestra director and amateur astronomer in Bath, was compiling a catalog of celestial objects that had not received much attention: double stars. One night in March, he noticed an object in the sky that he took at first for a comet. Successive sightings, however, revealed the object to be a planet, bringing the number of known planets in the Solar System to seven—the first such addition to the Solar System in recorded history (Hoskin, in Taton & Wilson 1995; 193). Herschel's discovery reinforced his scientific reputation, and he soon secured a position as a professional astronomer under the patronage of King George III.

Herschel's curiosity extended far beyond the observation of double stars. With the able assistance of his sister Caroline Herschel, he took up another neglected celestial object: nebulæ. Over the next two decades, Herschel published catalogs of some 2,500 nebulæ and clusters, grouped in eight categories distinguished by apparent luminosity and structure. Caroline, like her brother William, benefited from royal patronage; as the first woman to be so honored, she provided a brilliant example for other women to follow. William, in connection with his observations, calculated the relative motion of the Solar System among the stars, and used the result to build the first model of the structure of the Milky Way. The idea that the sidereal universe has a structure at all, which is to say, that stars are not randomly distributed, was advanced earlier by John Michell, in the first application of probabilistic reasoning in astronomy. The chance against an observed grouping of stars—the Pleiades constellation of six bright stars—being a random occurrence, Michell argued, is several billion to one. Some cause ruled the heavens, and for Michell, Newton's universal law of gravitation was a good candidate. The contributions of Michell, Herschel and others in the late eighteenth century effectively transformed the nature of astronomy, from a science based on the observation of the position of planets and comets in the Solar System, to one whose purview was the entire sidereal universe (North 2008).

Herschel's discoveries led to renewed questions concerning the stability of the Solar System and of the sidereal universe. Isaac Newton inaugurated such analyses with calculations of the secular perturbations of planetary orbits, which led him to believe that divine intervention was required in order to hold the Solar System together. As for the universe, Newton wondered how it could possibly be stable under universal gravitation. In correspondence with the classical scholar and cleric Richard Bentley, Newton agreed that in the case of an inhomogeneous distribution of matter, only a divine power could prevent the collapse of the universe.

The eighteenth century saw far-reaching progress in the sophistication of the tools of mathematical analysis, and by the turn of the nineteenth century, the question of the stability of the Solar System was ripe for reevaluation by Pierre-Simon de Laplace. The conclusion of Laplace's analysis was diametrically opposed to that of Newton, although he assumed with Newton that the dynamics of the Solar System depend only on the central force of gravitation. Laplace found the Solar System to be in a stable state, requiring nothing other than gravitation and inertial forces to keep the planets on course. Apparently, advanced mathematics had eliminated Newton's proof of the existence of God, although it seems unlikely that this was ever Laplace's intention.

Laplace, like many of his contemporaries in the Enlightenment, was a keen student of mathematical probability, a subject he took up in the 1770s. In 1812, he published a work—the *Analytical Theory of Probability*—which became a standard reference for nineteenth-century mathematical probability, and still defines classical probability theory. With Laplace, two grand ideas come together. First, there is the idea that nothing happens without a cause. The philosopher from Königsberg, Immanuel Kant, had written as much in his master work, *The Critique of Pure Reason* (1781): "All that happens is hypothetically necessary," and "Nothing happens through blind chance." Although Kant's theory of the foundations of knowledge in his *Critique* took inspiration from the success of Newtonian mechanics, Kant did not rely on this success to justify his negative view of chance events.

If nothing ever happens by chance, what is probability about? For Laplace, all that happens, happens because of natural law, even if we are unable to ascertain, in certain cases, how some events come about. This is where the laws of probability came into play. We can be certain that everything happens for a reason, Laplace wrote in the second edition of the *Théorie analytique*, because of the "regularity that astronomy shows us in the motion of the comets." The resounding success of celestial mechanics, in other words, extends to all phenomena, including, for Laplace, social phenomena.

It was no accident that Laplace picked out the motion of comets as his prime example of the descriptive power of Newtonian mechanics (Stigler 1999: 56). Cometary orbits can be quite unusual, unlike those of the planets. While the latter all follow slightly eccentric elliptical paths lying roughly in the same plane of the Solar System, comets are found in highly oblique planes, and undergo strong perturbations when in the vicinity of planets or the sun.

Following Herschel's discovery of Uranus, astronomers armed with new and more powerful telescopes were on the lookout for an eighth planet, which they thought would most likely be found in the space between Mars and Jupiter. The known planets all respected the empirical law of Titius-Bode, according to which the semi-major axis of the orbit a, measured in astronomical units, is given by the relation $a = 0.4 + 0.3 \times 2^n$, where $n = -\infty, 0, 1, \ldots$, such that the value of n is fixed by the sequential order of the planet. Yet, there was one exception: there seemed to be a planet missing between Mars ($n=2$) and Jupiter ($n=4$). An early form of the law was suggested in 1715, and most remarkably, the unexpected discovery of Uranus ($n=6$) served to confirm the law.

The hunt was on for the Solar System's eighth planet, and in 1801, an astronomer in Palermo, Giuseppe Piazzi thought he had sighted it, and he gave it a name: Ceres Ferdinandea, in honor of his patron, King Ferdinand IV of Naples and Sicily. The problem was that Piazzi was only able to obtain a few sightings before the object (which in public he called a comet, to be on the safe side) moved into the sun's glare, rendering impossible further observation. There was no analytical method available to determine the orbit of an object based on such slim data, which raised fears that Piazzi's comet would be lost. A young mathematician and astronomer in Braunschweig, Carl Friedrich Gauss learned of the situation, and on the basis on Piazzi's measurements, which covered only three degrees in a geocentric frame, he went to work on a new method for determining the orbit. Three months later, Gauss predicted the return of the object, roughly a year after its disappearance, to within a half a degree of its true position. Gauss' exploit was widely hailed, although Herschel refused to consider Ceres as the eighth planet, less because of its modest diameter, measured by him at 261 kilometers, than due to the inclination of its orbit with respect to the ecliptic, measured at thirty-four degrees. Instead, Herschel proposed Ceres as the first "asteroid."

It is likely that Gauss used the method of least squares on this occasion, in order to determine the "true" path of Ceres from the raw angular measurements of Piazzi. The method of least squares was first published in 1805 by Adrien-Marie Legendre in an appendix to a work on the determination of cometary orbits, but Gauss later claimed that he was in possession of the method already in 1795—a claim that irked Legendre considerably. Legendre had in fact given a straightforward and easily understood presentation of a method applicable to a wide variety of problems in which multiple observations are to be combined. He did not broach the question of the accuracy of the method, however.

Gauss, in his own presentation of the method of least squares, showed how probability entered the picture, thus affording a means of judging the accuracy of the method of least squares. In modern terms, Gauss sought the form of probability density and estimation method that lead to the arithmetic mean as estimate of the location parameter. To do so, he availed himself of the works of Laplace, including Laplace's form of Bayes' theorem, from which he derived what was later called a normal, or Gaussian distribution. Naturally, Laplace read Gauss' work with great interest. In doing so, he noticed its relevance for his own recent proof of an early version of the central limit theorem, generalizing De Moivre's approximation of the binomial theorem. Thanks to his reading of Gauss, Laplace now understood the connection between the central limit theorem and linear estimation. Unlike Gauss, Laplace justified the use of the method of least squares by its minimization of the posterior estimated error for a large number of trials. He pointed out that the normal error distribution leads to the arithmetic mean as the posterior mode, and is the only differentiable, symmetric distribution to do so. Laplace's new understanding underscored that if each measurement were considered to be the average of a large number of independent components, the distribution of such measurements would be normal.

By 1812, Laplace had brought together two currents of mathematical thinking: the combination of observations via aggregation of linear equations, on the one hand, and on the other, the assessment of uncertainty via mathematical probability. This intellectual and formal fusion opened up broad vistas for application of mathematical statistics, though appreciated at first only by those working in mathematics and the exact sciences of astronomy, geodesy, and physics. Even in these disciplines, which promoted the ideal of precision measurement, the use of statistical methods was largely limited to the domain of error correction via least-squares linear regression, to the neglect of uncertainty analysis.

FOURIER'S THEORY OF HEAT PROPAGATION

One physicist who quickly grasped the importance of Laplace's breakthrough was his former student at the École normale supérieure, Joseph Fourier. A former prefect, after Napoleon's downfall, Fourier was named director of the Bureau of Statistics of the Seine, and soon thereafter, he was elected to the Paris Academy of Sciences. From 1821 to 1829 the Bureau of Statistics published four volumes, containing both data and methods for their analysis. These methods were not new to mathematicians, but their publication aimed to benefit a broader public unfamiliar with higher mathematics. For example, one finds in a volume from 1826 (reprinted in Fourier's *Œuvres*), a technique for evaluating hypotheses with respect to a certain "standard of certainty." This proposes a formula for the degree of approximation of the calculated mean to the true mean, which differs by a factor of $\sqrt{2}$ from the modern definition of standard deviation.

Fourier also drew on Laplace's theory of probability in the course of his celebrated treatise, *The Analytical Theory of Heat* (1822). This work represented

FIGURE 5.1. Engraved portrait of Joseph Fourier (1768–1830). Photo by Stock Montage / Getty Images.

a clean break, not only with Biot's steady-state theory, but with the style of physical reasoning associated with Laplace and his disciples, issuing from celestial mechanics and Newtonian optics. Laplace expected all natural phenomena, from the microscopic scale to that of the universe, to obey a central force law on the model of Newton's universal law of gravitation, but admitting forces both attractive and repulsive (for more on this approach, see Chapter 4). Biot's account of heat diffusion was of just this sort. In its stead, Fourier favored a phenomenological approach, wherein a few choice facts are selected for mathematical generalization in the form of linear partial differential equations, boundary conditions are imposed, analytic solutions are sought and compared to observation. Fourier's theory of heat is distinguished by its generality, since

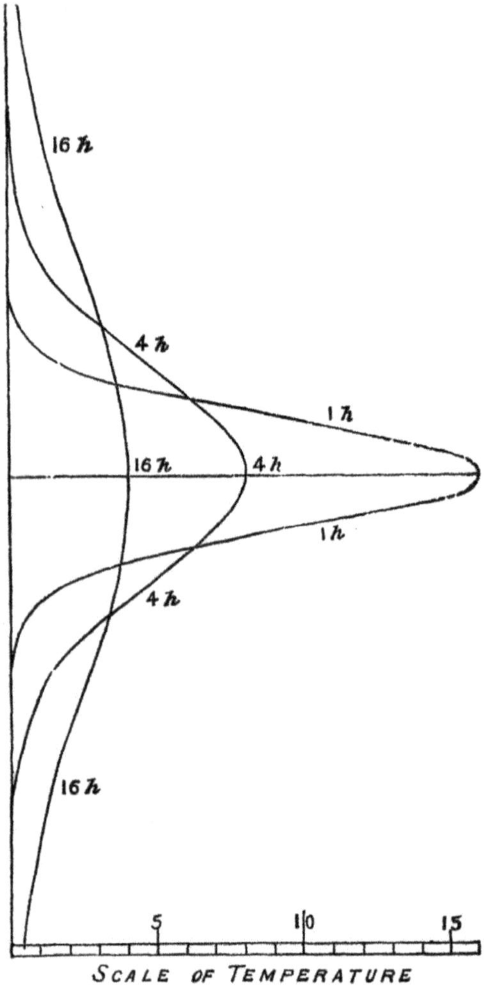

FIGURE 5.2. Heat distribution in an infinite solid at different moments of time after insertion of a hot layer, as calculated by Maxwell from Fourier's theory (Maxwell, *Theory of Heat*, 1871). Image courtesy of Scott A. Walter.

it contains no assumptions about the structure of matter. As a consequence, its scope of application is huge; in this sense, commentators consider Fourier's theory to constitute a veritable cosmology of heat, distinct from the mechanical cosmology of previous centuries.

It often happens in theoretical physics that the task of formulating a mathematical model of a given phenomenon turns out to be far simpler than solving the equations. This was true for Fourier in 1811, when he proposed an equation for heat flow in solids with a form recalling that of the wave equation:

$$\frac{K}{CD}\left(\frac{\partial^2 u}{\partial x^2}+\frac{\partial^2 u}{\partial y^2}+\frac{\partial^2 u}{\partial z^2}\right)=\frac{\partial u}{\partial t},$$

where u is the temperature of the medium, t is time, K is thermal conductivity, C is specific heat, and D is the density of the medium (Grattan-Guinness 1990). Finding solutions for solids of definite shape and temperature profiles can be a challenge, and here, Fourier proposed several new and powerful analytic methods, including dimensional analysis, and what was later known as Fourier analysis, involving harmonic distributions (on these methods, see Chapter 3). Fourier showed how the initial temperature distribution of a body can be modeled as pure surface effects, giving rise to a solution of the form $u(x,t)$, describable in terms of the linear sum of a series of sine and cosine waves, the coefficients of which are fixed by the geometry. This *Fourier series* expansion of the solution is inserted into the heat equation. Uniqueness of Fourier series guarantees that the coefficients on each side of the equation are equal, resulting in an ordinary differential equation, independent of time and easy to solve for the coefficients. While there are an infinity of coefficients in all but the most trivial cases, often calculation of the first few harmonics suffices to determine the temperature at any given position. Fourier's analytical method proved to be quite general, finding application in classes of physical and mathematical problems characterized by linear differential equations with constant coefficients, including electrostatics (§ Field Theory).

In addition to providing a powerful *analytical* framework for mathematical physics, Fourier's theory of heat flow proved to be suggestive in a *qualitative* sense for several branches of physics, beginning with electricity and magnetism. The invention of the Voltaic pile by Alessandro Volta at the turn of nineteenth century gave a huge boost to experimentation in this domain, which led, in turn, to the discovery of a series of surprising effects. Perhaps the most stunning of these came in 1820, when the Danish physicist Hans Christian Ørsted found that an electric current passing through a wire produces a circular magnetic force. Ørsted's discovery unified electricity and magnetism, which rapidly led to the foundation of electrodynamics by André-Marie Ampère, followed by Michael Faraday's theory of the electromagnetic field (Steinle 2016). In London, Humphry Davy found the conducting power of metal wires to be in inverse proportion to wire length and in direct proportion to cross-sectional area. He also found that conductivity varied inversely with wire temperature. In Berlin, Georg Simon Ohm extended Davy's investigations, studying the effect of varying the driving-power of a Voltaic battery, or what was later known as voltage. In doing so, Ohm supposed that electricity propagates along a wire from one particle to the next. He further assumed that the magnitude of this flow is proportional to the voltage difference between adjacent particles, just as in Fourier's theory, where the flow of heat between two points depends on the temperature differential. Ohm's experiments led him to advance his eponymous law (1827) relating electric current to voltage and wire resistance in a circuit.

ADOLPHE QUETELET'S SOCIAL PHYSICS

The science of probability was an integral part of Fourier's cosmology, a fact reflected by the faith he placed in the validity of the normal distribution for representing not just errors, but natural frequencies. For example, Fourier considered a physical model in which heat is concentrated in one point of an infinite, one-dimensional conductor. Based on his heat equation, Fourier calculated the resulting distribution, which was just the normal distribution. From an epistemological standpoint, since both the heat equation and Laplace's binomial derivation of the normal distribution lead to the same result, both are equally confirmed by experimental observation in this instance, making a preference for one explanation over the other a matter of taste.

The success of the probabilistic approach was suggestive to Fourier, who sought a molecular representation of the nature of heat. "Molecules separated from one another," he wrote in 1822, "communicate their rays of heat reciprocally across airless space, just as shining bodies transmit their light." Heat dispersion, according to Fourier's model, is the result of rays transmitted by contiguous molecules, in conformity with his heat equation. Heat flow, he observed, "then results from an infinite multitude of actions whose effects are added." Such a view of molecular action naturally lent itself to a great variety of physical phenomena, and not just those concerning heat dispersion. Likewise for the Gaussian law of error, which Fourier suggested was "imprinted in all of nature—a pre-existent element of universal order" (Fourier 1819, cited by Porter 1986: 99).

Fourier's mathematical approach to physics deeply impressed scientists in France and throughout Europe. One of these scientists was a Belgian named Adolphe Quetelet. Born and educated in Ghent when this city was a part of France, Quetelet had great admiration for the accomplishments of the French mathematicians Laplace, Lacroix, Fourier and Poisson. Soon after he obtained a PhD in mathematics, Quetelet was sent to the Royal Observatory in Paris for three months to obtain instruction in the finer points of astronomy. When he returned to Brussels in 1823, he brought with him an appreciation—and a certain enthusiasm—for the science of statistics.

In Paris, Fourier had facilitated the application of the latest statistical techniques to the data collected by the Bureau of Statistics, but he was not otherwise concerned with a probabilistic understanding of human society. The science of both Fourier and Laplace exercised the mind of young Quetelet, who sought to extend these methods to human affairs. In doing so, he latched onto an analogy promoted by Laplace, who claimed that just as the law of universal gravitation subtends order in the physical realm, the eternal laws of truth, justice, and humanity maintain order in the civil realm. Laplace had investigated these eternal laws, if only in a limited sense, when he used the laws of probability to ascertain how judicial error depends on the number of jurors.

Quetelet believed statistical analysis would identify the laws governing civil order, and he aspired to become the "Newton of this other celestial mechanics," thus, the master of the mechanics underlying the storm of social phenomena (cited by Porter, 1986: 104). One implication of Quetelet's understanding of the laws of social phenomena, an inference drawn by many in this emergent field, was that statisticians required training in advanced mathematics. Another implication was that any measurable entity related to humans and human activity was susceptible to statistical analysis, whether it be a physical quantity like body weight, or a behavioral statistic, like the incidence of violent crime.

Statistics on crime were published in France from 1827, and in the 1830s, the statistical movement grew rapidly in France and Britain. The British Board of Trade published economic statistics from 1832, and five years later the newly created General Register Office began collecting census and mortality records. Public hygiene was a priority in France and Britain, where statistical societies were founded in major cities and took part in campaigns promoting public education, sanitation, and poor law reform. No longer reserved to specialists, statistics had entered the sphere of public discourse.

In the midst of the institutional and popular rise of statistics, Quetelet published his first major work, *On Man and the Development of His Faculties, or an Essay on Social Physics* (1835). One notion he introduced found great resonance among his readers: the idea of the "homme moyen" or the average man (Stigler 1986: 169). (Since Quetelet did not write about the "personne moyenne," though he might have, this translation reflects the original usage and contemporary understanding of his writings.) What is an average man? For Quetelet, he was not a human being but rather a mathematical fiction meant to represent the set of arithmetic means associated with the statistics of a homogeneous population of men. Consequently, the average man would have different characteristics for different populations.

A few years later, Quetelet made further inroads toward a generalization of statistical reasoning by considering a normal distribution to be the "natural" distribution, much like Fourier's consideration of uni-dimensional heat flow, mentioned above. Already in 1818, the German astronomer Friedrich Bessel examined the departures from the normal distribution of 60,000 measurements of 3,222 stars, and suggested sources of variation other than simple measurement error. In other words, Bessel compared the normal distribution with observed stellar position to judge the nature of the variations in positional reports for a given star. Bessel might have taken the opportunity to evaluate the method of least squares, but he did not do so. None of his contemporaries did so, either.

TABLE 5.1. The height of French conscription candidates, from Quetelet (1844).

	Number of males		
Male height	*Measured*	*Calculated*	*Difference*
Below $1^{m}.570$	28,620	26,345	+2,275
$1^{m}.570 - 1^{m}.597$	11,580	13,182	–1602
$1^{m}.597 - 1^{m}.624$	13,990	14,502	–512
$1^{m}.624 - 1^{m}.651$	14,410	13,982	+428
$1^{m}.651 - 1^{m}.678$	11,410	11,803	–393
$1^{m}.678 - 1^{m}.705$	8,780	8,725	+55
$1^{m}.705 - 1^{m}.732$	5,550	5,527	+3
$1^{m}.732 - 1^{m}.759$	3,190	3,187	+3
Above $1^{m}.759$	2,490	2,645	–155

Perhaps the most obvious reason for this state of affairs is that there was simply no alternative at hand to the method of least squares. The idea is nonetheless present, suggesting that in some cases, the variation of an observed quantity may be compared fruitfully to a normal distribution.

As for Quetelet, even though he was a founder and director of the Brussels Observatory, his creative force was directed less to astronomy than to social physics. He also had great faith in the power of the central limit theorem. For instance, much as Bessel had wondered how to interpret a non-normal distribution of the observed position of a star, Quetelet expected the height of adult males in a given population to be normally distributed. When the statistics showed a distribution of height in a given population that was not normal, Quetelet inferred that there was a reason for this: an unknown cause must have intervened, resulting in a deviation from the normal distribution of height.

Quetelet noticed, in particular, that the heights of French conscription candidates, while featuring a single apex as in a normal distribution, differed from such a standard distribution because it was neither symmetric nor smooth. What could explain such an anomaly? The number of hypotheses one could introduce was nearly limitless, but Quetelet seized upon the following: bad data. His height statistics, Quetelet inferred, were marred by fraud. Quetelet knew, of course, that those 20-year-old Frenchmen shorter than 157 centimeters were exempted from conscription. For those preferring not to serve—presumably a large sub-population among French 20-year-old males—and who happened to be only slightly taller than the cutoff, the incentive to persuade the authorities to shave a centimeter or two off the measured height was apparent. As for the authorities, Quetelet reasoned that they would be willing to play along in borderline cases, since the individuals concerned, being of modest stature, would be replaced by others endowed with a "more advantageous height."[1]

In Quetelet's hands, the normal distribution became a means by which to evaluate a population's homogeneity. Some contemporaries realized this was an unjustified point of view and not just from a mathematical standpoint. Quetelet's other ideas, including that of the average man, were also subjected to criticism. His critics, however, were drowned out by the general acclaim accorded social physics. Taking into account Quetelet's background and the rising authority at this time of astronomers and physicists in public discourse, the popularity of his ideas is not difficult to understand. In a period of great socioeconomic and political upheaval, accompanied by a tidal wave of social statistics, what he offered to social reformers was incredibly seductive: a rational foundation for shaping society. While his misconceptions were many, Quetelet's goal of ascertaining the principles of social order through mathematical analysis was both laudable and in phase with the progressive spirit of the mid-nineteenth century.

Although Quetelet's writings on probability were widely translated, read and studied, their long-term impact was negligible, as applications of error theory made no significant gains in the social sciences until the twentieth century. Error theory remained important in astronomy and physics, the areas from which it had emerged, but biology, meteorology, geology, and empirical economics went largely untouched by it. Several reasons for this might be considered, but among them one stands out. In celestial mechanics, and Newtonian dynamics in general, one relies on equations of motion as a guide to studying the object under investigation, which thereby gains a certain objectivity independent of measurement and probability. When Gauss calculated the orbit of Ceres, he fit measured positions to an elliptical orbit. While the measured positions were assumed to be slightly in error, the true position of Ceres was nevertheless assumed to coincide with an elliptical orbit (barring eventual perturbations caused by the gravitational attraction of nearby planets). The same could not be said of the objects accruing to other disciplines. For example, there is no law of human development which would affirm the true height of 20-year-old Frenchmen to be, say, the median height. Likewise, longevity was not normally distributed, as humans from favorable socio-economic backgrounds live longer, on average, than those from the poorer classes. Quetelet thought he knew why: the rich enjoyed the benefits of an "enlightened will . . . [and] habits of propriety, of temperance, of passions excited less frequently and variations less sudden in their manner of existence" (cited by Porter 1986: 103).

MACHINE EFFICIENCY AND MATHEMATICAL ECONOMICS

Other social theorists suggested, more sensibly, that wealth per se granted privileged access to adequate nutrition and clean water, which resulted in greater longevity on average. Yet, even the founder of Marxism, Karl Marx found

Quetelet's concept of the average man led to a uniform standard of labor. This then provided a precise foundation for the labor theory of value, which correlated the value of goods with the labor required to produce them (Porter 1986: 66). In the 1870s, the labor theory of value was supplanted by the marginal theory of value, in which wages vary. The first mathematical theory of economics, Augustin Cournot's *Researches into the Mathematical Theory of Wealth* (1838), borrowed heavily from the laws of physics, which still lacked a statistical component. Just as Quetelet believed the civil realm to be ruled by laws like those that ruled the motion of the planets, so did Cournot model his rational economics on rational mechanics. Since the productivity of a machine could be calculated from the laws of physics based on the energy extracted in the form of work, one could also measure the productivity of an enterprise by its profit report. In both cases, maximization of productivity was the objective. The French-trained Swiss economist Léon Walras even took Cournot's mechanics-based approach a step further by introducing the notion of equilibrium. In Walras' scheme, prices of goods interact in a perfect market based on demand curves, rationally determined by the relation between the utility of goods and their abundance. The calculation of marginal utility he introduced was deterministic and based on infinitesimal calculus.

The idea of extracting work from machines was one that emerged with the onset of the First Industrial Revolution. The Watt steam engine and its successors provided power to manufacturing sites, and from the turn of the nineteenth century reciprocating steam engines drove the locomotives that delivered raw materials to these sites before moving finished goods to markets. From the 1820s, works on the theory of machines and industrial mechanics in France drew on a century-old engineering tradition of extracting power from water, wind, and steam. In this context, the efficiency of hydraulic engines was measured by their capacity to restore a volume of the motive agent to its source, which is to say, a measure of the original head of water. A hydraulic engine capable of restoring the original head in its entirety was, by definition, a perpetual-motion machine.

A comparable characterization of the efficiency of a steam engine remained elusive until the French engineer, Sadi Carnot, gave the problem some thought. Carnot, the son of Napoleon's minister Lazare Carnot, was trained at the École polytechnique. Drawing on a number of intellectual sources, including Fourier, Sadi Carnot realized that an analogy between hydraulic engines and heat engines (such as steam engines) was at hand. In this analogy, the impossibility of perpetual motion, expressed for hydraulic engines as the impossibility of restoring entirely the original head, has its counterpart in the ideal heat engine. Such an engine extracts work, though not from a difference of *head*, but rather from a finite *temperature* gradient between two bodies; conversely, it produces a temperature difference in the two bodies when work is supplied. This process being reversible, it defines the maximum efficiency of any heat engine. Also, if the two bodies are at the same temperature, the ideal heat engine produces no work. Finally, unless work is applied, heat always flows "downhill" from a warmer body to a colder one and never "uphill," from a colder body to a warmer one. Carnot's principle expresses succinctly the second law of thermodynamics, elaborated three decades later by William Thomson and Rudolf Clausius, and in fact, Carnot's *Reflections on the Motive Power of Fire* (1824) was long ignored in his home country and also abroad, with one important exception.

In 1834, Émile Clapeyron, a fellow graduate of the École polytechnique, gave Carnot's ideas mathematical as well as graphical expression. Quickly translated into English and German, Clapeyron's work brought Carnot's ideas to the attention of Thomson and others by the 1840s, when it inspired groundbreaking advances in thermal physics and field theory. Carnot's concept of an ideal heat engine was likely ignored at first in part due to its novelty, and yet in one respect it was quite traditional. Carnot simply pushed the analogy between hydraulic and heat engines too far because he thought of heat as a substance—*caloric*—a subtle form of matter akin to a highly elastic fluid. The caloric theory arose during the eighteenth century and was promoted by the great chemist Antoine Lavoisier.

A tax collector during the *Ancien Régime*, Lavoisier fell victim to the violence of the French Revolution, but his wife survived, and wed another man of science, the American expatriate Benjamin Thompson, Graf von Rumford. Rumford presented an argument tending at once to disprove the caloric theory of heat, and to affirm the rival kinetic theory. At the arsenal in Munich, where he served as superintendent, Rumford observed that the surface of a cannon became quite hot during a boring operation. This gave him the idea of a

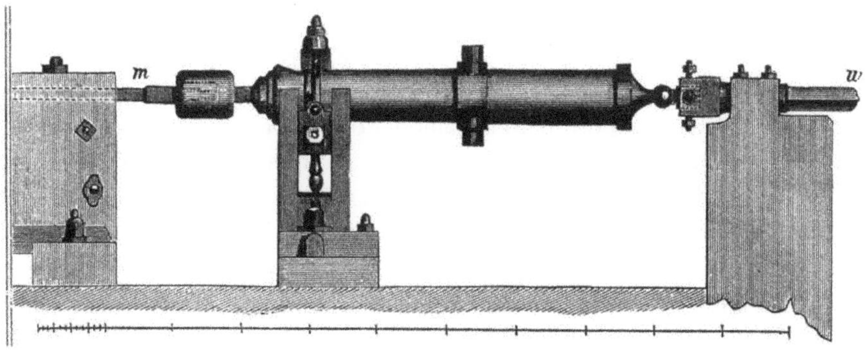

FIGURE 5.3. Rumford's schematic (from Phil. Trans., 1798) of an experiment, whereby heat is generated from the friction between a blunt steel head attached to a fixed bar (m) and a brass cannon driven to rotate along its axis via a larger bar (w). Image courtesy of Scott A. Walter.

"beautiful experiment," in which he replaced the boring head with a flat piece of steel pressed against one end of a rotating brass cannon, and immersed the contact surface in a water bath. After two and a half hours of frictional contact, with the cannon rotating at 32 rpm, Rumford reported that the water began to boil—to the astonishment of onlookers, there being no fire. The weight of the metal powder resulting from the experiment was less than 270 grams, and Rumford considered improbable the idea that caloric liberated from this minute quantity of material could bring about such a significant increase in temperature of the water bath. He thus concluded that the heat generated in this case was not due to a material substance but rather to the motion of the cannon, which is to say, it was kinetic.

THERMODYNAMICS AND THE KINETIC THEORY OF GASES

Rumford went on to embrace theories of heat and light as wave phenomena, while in England, John Herapath developed an image of heat as the motion of atoms (Brush 1983). The absolute temperature of a gas, Herapath argued, is a direct function of molecular velocity (instead of the square of velocity, as Daniel Bernoulli had it), where molecules are modeled as elastic spheres. By the 1820s, the caloric theory was in decline, but since Herapath's theory itself was plagued with inconsistencies, it, too, gained few followers. One of these, however, was James Joule, who compared the atoms of a gas to high-speed projectiles, flying in all directions. Whereas Carnot's ideal heat engine posited the inter-convertibility of heat and mechanical work, Joule, the son of a brewer and former student of the atomist John Dalton, performed an experiment to determine quantitatively the mechanical equivalent of heat. Using a paddle wheel immersed in an insulated jar of water and attached to a mechanism driven by two falling weights, he determined the precise amount of mechanical power required to raise the temperature of a pound of water by 1°F. His experiments led him to assert, contrary to Carnot, that whenever work is produced by heat flow, a quantity of heat is consumed in proportion to the amount of work extracted.

Joule's experiment impressed the German physicist Hermann Helmholtz, who agreed with the view that heat is a form of motion. As such, Helmholtz understood that the law-like conversion of heat to work, and vice-versa, meant that the law of conservation of mechanical *vis viva* could be extended to a universal conservation of energy, or in Helmholtz's terminology, *Kraft*. Electricity and magnetism were subject to the principle, as well, as Helmholtz observed that the passage of electric current produces heat. In a bold generalization from extant knowledge, he predicted that all physical interactions, including chemical and organic processes, would likewise verify the principle of conservation of energy.

When Helmholtz unified all known forces under the principle of energy conservation, he discarded both the caloric theory and the notion, promoted earlier by Ampère, of heat propagation by means of wave motion, preferring instead to embrace a theory of matter in motion. Three years later, Rudolf Clausius gave the principle of energy conservation mathematical expression. A former student of Dirichlet, and of the historian Leopold von Ranke, Clausius was teaching physics in Berlin when he came upon Clapeyron's paper, which impressed him a great deal, although he found it lacking in rigor. In particular, Clausius (following Joule and Thomson) was not convinced that work could be extracted from a heat engine without some concomitant consumption. Abandoning Carnot's assumption that no heat is lost in a cyclic process, he managed to reconcile the views of Carnot and Joule by affirming the latter's position on the equivalence between heat and mechanical work. Clausius thus held that when an ideal heat engine produces work, it consumes a certain quantity of heat. Moreover, another quantity of heat is lost in passing from a warmer body to a colder one, and both of these quantities are proportional to the amount of work extracted. Today we recognize in these two principles the first and second law of thermodynamics, respectively.

In 1854, Clausius sharpened the expression of the second law of thermodynamics and fundamentally transformed both the science of thermodynamics as well as scientific understanding of the physical world. He noticed that Carnot's principle expressed the motive power of fire as a difference of temperature, and not of heat. He also noticed that Carnot considered two transformations of heat: (1) the lossless heat transfer from a warmer body to a colder one, and (2) the conversion of heat from the warmer body into work. The two transformations are equivalent, Clausius claimed, provided that the heat cycle is reversible.

This equivalence of transformations led Clausius to affirm the equivalence of corresponding "equivalence-values" of such transformations. From the assumption that the product of equivalence-values of two transformations is equal to the equivalence-value of the product of their transformations, and that the total equivalence-value of transformation of a reversible cycle of a bi-thermal heat engine is zero, Clausius derived the expression

$$Q_1 f(T_1) + Q_2 f(T_2) = 0,$$

where Q and T are quantities of heat and temperature. Clausius knew of Thomson's definition of absolute temperature, according to which

$$\frac{Q_1}{T_1} + \frac{Q_2}{T_2} = 0,$$

so he defined $f(T)$ to be proportional to $1/T$ and went on to consider reversible cyclic processes of a general form. Writing δQ for the net heat flow from a body at temperature T, this leads to:

$$\oint \frac{\delta Q}{T} = 0.$$

For a transformation realized by an *irreversible* cycle of a heat engine, the value of the integral would necessarily differ from zero. Clausius chose to assign a negative value to the integral in this case, where for any closed system,

$$\oint \frac{\delta Q}{T} \leq 0.$$

In the reversible, quasi-static case, considering that the working fluid takes on the temperatures of the bodies exchanging heat, Clausius saw that the transformation value is a property of the fluid, and since it vanishes on completion of a cycle, it is a complete differential. From the Greek word for transformation (τϱοπή), Clausius introduced the term "entropy," and taking the universe for the object system, he re-expressed the fundamental laws of thermodynamics as follows:

1. The energy of the universe is constant.
2. The entropy of the universe tends to a maximum.

Along with Joule and Helmholtz, Clausius felt the equivalence of heat and mechanical work to be consistent with the hypothesis that heat is due to molecular motion (Brush 1983). Starting in 1857, he introduced the kinetic theory of gases to thermodynamics with an aptly entitled paper, "On the kind of motion we call heat." Objections to this theory, which equated the temperature of a gas with the mean value of the squared velocity of its constituent molecules, led Clausius to adopt a more sophisticated approach, relying on the statistical concept of a "mean free path," defined as the average distance covered by a molecule before it strikes another molecule. The approach was bold, inasmuch as he had no idea how molecules interact, but Clausius made a few assumptions that seemed reasonable and formulated the first approach to molecular motion as a stochastic process.

Calculation of the mean free path

A key assumption made by Clausius in his derivation of the mean free path was that of the "sphere of action." A gas molecule moves unperturbed in a straight line until it enters the sphere of action of another molecule, where this sphere of action is fixed by the value of a radius ρ extending from the center of the molecule. To find the probability that a point passes unperturbed through a gaseous layer of thickness x, Clausius noted that if the probability is a for unit thickness, then for thickness 2 the probability is a^2, and the probability $P(x)$ that a molecule traverses without collision through a layer of thickness x is

$$P(x) = e^{-\alpha x},$$

where α is a positive constant. Assuming further that collision probability is just the ratio of volume swept out by the spheres of action to the volume of the gas layer of molecules with mean separation λ, Clausius let $\alpha=\pi\rho^2/\lambda^3$, such that, for small x,

$$P(x) = e^{-\frac{\pi\rho^2 x}{\lambda^3}} \approx 1 - \frac{\pi\rho^2 x}{\lambda^3}.$$

If we denote the density per unit volume as n such that $n = 1/\lambda^3$, then we have

$$-P'(0) = \alpha = \pi\rho^2 n.$$

The mean free path for the case of all molecules save one at rest, denoted ℓ_∞, is then

$$\ell_\infty = -\int_0^\infty xP'(x)dx = \int_0^\infty P(x)dx = \frac{1}{\alpha} = \frac{1}{\pi\rho^2 n}.$$

Clausius reasoned that the difference in mean free path for this case and that in which all molecules move with like speed is $\ell = 3\ell_\infty/4$, such that the mean free path of the latter—more realistic—case is just

$$\ell = \frac{3}{4\pi\rho^2 n}.$$

By assuming a certain value for the radius of the sphere of action, Clausius found that his mean-free-path approach explained rather well the fact that smoke clouds disperse slowly when prevailing winds are gentle.

One of Clausius' early readers was the young James Clerk Maxwell (Flood, McCartney, and Whitaker 2014), who was familiar with the rudiments of probability theory, having read a review of Quetelet's work published by William Herschel's son John in 1850. Maxwell came up with a slightly different expression for the mean free path length, with only the slightest of justification: $\ell = \ell\infty/\sqrt{2}$. Clausius was nonplussed, and published a correction, but in fact Maxwell's formula follows from more rigorous arguments than the one he offered at the time, as Clausius later recognized.

Maxwell also discerned a way to determine the equilibrium distribution of molecular velocity, $f(v)$, such that the average number of molecules with speeds between v and $v+dv$ is $f(v)dv$. Assuming statistical independence of velocity components and isotropy of the distribution, Maxwell concluded that

$$f(v) = N\frac{4}{\alpha^3\sqrt{\pi}}v^2 e^{-\frac{v^2}{\alpha^2}},$$

FIGURE 5.4. James Clerk Maxwell (1831–79). Photo by SSPL / Getty Images.

where α^2 is inversely proportional to absolute temperature, and N is the number of molecules. One consequence of Maxwell's theory gave him pause: gas viscosity is independent of density. This result appeared highly counter-intuitive, but subsequent experiments, performed with the help of his wife Katherine, confirmed the mathematical result and did much to recommend kinetic theory to physicists.

Maxwell was not satisfied with his derivation of the velocity distribution law, and both he and the Austrian physicist Ludwig Boltzmann tried to improve upon it. In a celebrated paper of 1866, Maxwell developed the theory of transport processes in gases. This concerned viscosity, heat conduction, and diffusion, based on a model in which molecules are represented by repulsive point-centers of force. Solution of the theory's equations was beyond reach, save in the special case of a force varying as the inverse fifth power of distance. In this case, Maxwell recovered the Navier-Stokes equation for motion of viscous gases, supplying quantitative predictions relating thermal conductivity, diffusion rate, and viscosity. Later experiments, however, tended to disconfirm the temperature dependence of viscosity for the $1/r^5$ case.

Soon after Maxwell's second kinetic theory appeared, Boltzmann generalized it to molecules with internal degrees of freedom; this allowed for influence by external forces, such as gravity (Brush 1983). The new distribution function thus depended not only on velocity, but on potential energy, such that the relative probability of finding a molecule with potential energy V at a given position is expressed in the form $e^{-V/kT}$, where k is Boltzmann's constant, and T is absolute temperature. The combination of kinetic and potential energy is given by

$$E = \sum \frac{1}{2} mv^2 + \sum V.$$

and the probability of any molecular state with total energy E in a gas, liquid, or solid, is $e^{-E/kT}$.

To demonstrate the uniqueness of Maxwell's probability distribution, Boltzmann drew on Maxwell's method of determining collision frequency, from which he derived an equation that determines the temporal evolution of the probability distribution f from an initial value. Boltzmann assumed the probability that two molecules with velocities v_1 and v_2 collide is equal to the product of the individual probabilities:

$$F(v_1, v_2) = f(v_1)f(v_2).$$

Every collision changes the value of f for four values: two before, and two after the collision. Since the probability of each collision depends on f, the temporal evolution of the probability distribution is described by an integro-differential equation of the form:

$$\frac{\partial f}{\partial t} = \int (f(v_1')f(v_2') - f(v_1)f(v_2))\, g(v_1, v_2, v_1', v_2'),$$

where the primes indicate post-collision velocities, and $g(v_1, v_2, v_1', v_2')$ accounts for inter-molecular forces. Additional terms may be introduced to account for temperature gradients, fluid velocity and external forces, leading to what was later known as Boltzmann's equation. When Boltzmann examined the expression

$$H = \int f \ln f\, d^3 v,$$

he found that, as a consequence of his equation for temporal evolution of f, the value of H decreases strictly with time for all distributions except for that of Maxwell; this is Boltzmann's "H-theorem." When f is taken to be Maxwell's distribution, Boltzmann noted, the value of $-H$ is just the entropy of Clausius, while for other, non-Maxwellian distributions, the function can be regarded as describing non-equilibrium states.

Boltzmann's H-theorem thus gave rise to a probabilistic interpretation of entropy, and thus a statistical interpretation of the second law of thermodynamics. In the 1890s, Boltzmann argued that an irreversible approach to the state of equilibrium, as an entropy-increasing process, is just a transition from microstates with lower probability to microstates with higher probability, and that consequently entropy is a measure of probability. For a macrostate probability W, the entropy S of the macrostate is then described as

$$S = k \ln W.$$

The kinetic theory briefly fell out of favor in the 1890s, despite its great promise as a solid foundation for the theory of transport processes and its success in describing phase transitions in Van der Waals' theory of intermolecular interactions. It also presented significant mathematical and conceptual challenges that discouraged some theorists until after the turn of the century. The outlook for kinetic theory changed for the better largely due to Max Planck's theory of black-body radiation, which relied on Boltzmann's statistical expression of entropy.

FIELD THEORY

Another theory of physics that went through an extended period of general neglect on the part of the scientific community was the field theory of Faraday and Maxwell (Darrigol 2000). Michael Faraday's field theory of electricity and magnetism, elaborated in the 1830s and 1840s, opened up new vistas for physics, but gained few adepts until the 1850s. One crucial exception was the young William Thomson, who in 1841, just before starting university, showed via a formal analogy how Fourier's theory of heat propagation could be successfully adapted to problems of electric potential. Thomson modified slightly Fourier's treatment of heat sources, by showing that any isothermal surface has an equivalent expression in terms of point sources. Earlier writers, including Gauss, Laplace, and Poisson, had considered the potential *V*:

$$V = \int\int\int \frac{\rho}{r} d\tau,$$

where ρ is matter density (or electric density), r is distance from the attracting center to the attracted point, and $d\tau$ the volume element, to express the gravitational (or electrical) force F at the attracted point as

$$F = -\nabla V,$$

The potential V further satisfies the equation

$$\nabla \cdot \nabla V = \nabla V = -4\pi\rho.$$

Compare this to Fourier's equations (§ Fourier's Theory):

$$F = -K\nabla V,$$

and

$$\frac{K}{CD}\nabla \cdot \nabla V = \frac{K}{CD}\Delta V = \frac{\partial V}{\partial t}.$$

In Thomson's analogy, the point sources of heat, which according to Fourier's heat equation result in a temperature profile proportional to the reciprocal of distance from the source, correspond to electric charges. He showed further that the equipotential surfaces generated in space by charged conductors are analogous to the isothermal surfaces of an infinite solid body through which heat propagates.

Thomson's analogy is particularly striking when we recall that he was still a teenager when he first formulated it in 1841. As it happened, the important theorems in his paper were already known. Yet, even though C. F. Gauss, Michel Chasles, and George Green proved these earlier, none of these mathematicians had noticed the analogy between heat conduction and electrostatics. Thomson's physical analogy worked so well that he naturally sought others, in both physics and mathematics. Upon gaining greater familiarity with Faraday's conception of electrostatics in terms of "lines of force," Thomson expanded his earlier analogy from Fourier heat theory to electrostatics. He noted, in particular, that Faraday's tubes of lines connecting conductors, which were everywhere tangent to the electric force, are formal correlates to Fourier's lines of heat flow, everywhere orthogonal to isothermal surfaces. Thomson also noticed that the "tension" measured by Faraday in his experiments on inductive capacity was a physical realization of Green's "potential." This was described mathematically in 1828, but Thomson was the first to define it in an operational sense. Faraday, for his part, was intrigued by Thomson's analogies, and from the mid-1840s the two men struck up a mutually stimulating correspondence.

The meeting of Thomson and Faraday would soon shake the world, but not as a direct result of tabletop experiments or theoretical analogies; rather this came through their engagement with a frustrating problem in cable telegraphy. In 1851, a submarine cable already connected Dover and Calais and plans were soon underway for a transatlantic cable. After this was first realized in 1858, poor cable design and mishandling soon rendered it inoperative (Gordon 2002). While early submarine cables promised significant commercial and military advantages, scientists, too, stood to gain from the new technology. In an effort to improve the precision of longitude measurement, the Astronomer Royal, George Biddell Airy, collected data on the time lag of signals transmitted between Greenwich and the observatories in Edinburgh (540 km) and Brussels (313 km). The data showed, to Airy's bewilderment, that despite the greater distance, Edinburgh's signal time lag was nearly imperceptible, while for Brussels it came to about one tenth of a second. The exact cause of this difference was unclear, but the culprit seemed obvious: the telegraph cable to Brussels was mostly subground or submarine, while the Greenwich-Edinburgh line was mostly airborne.

The anomaly of differential time lag in telegraphic signals was brought to Faraday's attention, and he promptly explained it during an evening lecture at the Royal Institution, held on 20 January, 1854 (Hunt 2021: 22). The signal retardation, Faraday observed, arose from the gutta-percha sheathing used to insulate the copper wires. Current conduction, in his view, did not occur like water running through a pipe, and in fact, Faraday wholly rejected the image of a flow of electricity. Rather, the real seat of electromagnetic phenomena resided in the space *around* the wire. The circular magnetic force discovered by Ørsted (§ Fourier's Theory) constituted a "field" in space; thus, for Faraday, the electric charges were secondary phenomena, while the field itself was primary. In the case of an insulated copper wire, the existence of current required the induction of a state of strain in the surrounding dielectric, i.e., the wire's sheathing. According to Faraday's theory of the electromagnetic field, the observed time lag in signal propagation along copper wires could thus be explained qualitatively by the difference in inductive capacity between the thick sheathing of buried and submarine cables as compared with the thin sheathing used for airborne cables. Faraday's explanation provided a compelling argument in favor of his field theory, which most physicists had ignored in previous decades. Prior to the laying of long-distance telegraph cables, there was no real evidence for the propagation of electrical actions in time—electrical effects all seemed instantaneous, or nearly so. In other words, Faraday's field theory of dielectrics, when he first advanced it in 1842, was a solution in search of a real problem.

Faraday's Royal Institution lecture naturally attracted Thomson's attention. In a letter to George Gabriel Stokes, Thomson, recalling his earlier adaptation of Fourier's equations to electrostatics, now cast Faraday's notion of specific inductive capacity in the form of a second-order differential equation:

$$RC\frac{\partial V}{\partial t}=\frac{\partial^2 V}{\partial x^2},$$

where R is the resistance per unit length, C stands for capacitance, V for the potential, and x is the cable length. Accordingly, Thomson predicted that the time required to reach a given percentage of maximum current intensity at the far end of a cable is proportional to the square of cable length l, or RCl^2. Tests of time lag for relatively short lengths of insulated cable could thus be used to predict with great precision the signal time lag for very long cables, such as the 3,200 kilometer transatlantic cable. Using the data collected by Airy for the Greenwich-Brussels cable, Thomson was able to model the time lag for a transatlantic version, assuming the same cable design, but with a greatly extended length. The result was rather bad news for the cable company, as Thomson's model predicted a 10-second time lag for the proposed cable which, when combined with a concomitant signal deformation, drove the data-transmission rate so low that a transatlantic cable telegraph system would not be economically viable.

The general enthusiasm for the great submarine cable was not easily quashed. An intrepid surgeon from Brighton contested Thomson's "squared-length" law of signal retardation and presented measurements that seemed to support his contention. Thomson, however, was unperturbed, and he explained to the surgeon how some of his measurements, on the contrary, confirmed the squared-length law, while others failed to take into account either the finer points of his theory or those of electrical measurement. The upshot of the exchange was that Thomson became a director of the Atlantic Telegraph Company.

Thomson's work on submarine cable technology made him a wealthy man, in spite of the utter failure of the cable laid in 1858. He attributed this failure to manufacturing and production errors, particularly with respect to the gutta-percha sheathing. In light of Thomson's report, plans were made for a new-and-improved submarine cable between Ireland and Newfoundland, after

FIGURE 5.5. Paying out the transatlantic submarine telegraph cable in 1865. Photo by Kean Collection / Getty Images.

enlisting investors willing to finance another attempt. Although the American Civil War of 1861–5 delayed realization of this project, it was completed in the summer of 1866. The success of this venture brought with it a knighthood for Thomson. He was subsequently elevated to the peerage as Baron Kelvin of Largs, the first scientist to be so honored, although his political activity in West Scotland had, in fact, weighed more heavily than his scientific accomplishments.

Thomson's biographers have noted the symbolic power the transatlantic cable exercised on British minds in the late nineteenth century. This technological exploit served to confirm a seductive narrative of human progress highlighting humanity's power to subdue and control the mighty forces of nature. The Old World of Europe was now connected—"instantaneously," as Thomson said—to the new one of the Americas. This added significantly to the reach of the electric grid, giving rise to the image of a world connected by telegraph, railway, and ocean liners in an "electric nerve" system of a global body (*The Engineer*, 1857, quoted by Smith and Wise 1989: 651).

The poet Rudyard Kipling expanded on the notion of a global body in his poem "The Deep-Sea Cables" (1893), which speaks to the spiritual reach of the newest technological wonder:

The wrecks dissolve above us; their dust drops down from afar –
Down to the dark, to the utter dark, where the blind white sea-snakes are.
There is no sound, no echo of sound, in the deserts of the deep,
Or the great grey level plains of ooze where the shell-burred cables creep.
Here in the womb of the world – here on the tie-ribs of earth
Words, and the words of men, flicker and flutter and beat –
Warning, sorrow, and gain, salutation and mirth –
For a Power troubles the Still that has neither voice nor feet.
They have wakened the timeless Things: they have killed their father Time:
Joining hands in the gloom, a league from the last of the sun.
Hush! Men talk to-day o'er the waste of the ultimate slime.
And a new Word runs between: whispering, "Let us be one!"

Kipling's affirmation that cable telegraphy killed time involves a measure of poetic license. According to Thomson's telegraphy equation, confirmed by the message transmission rate of eight words per minute achieved by the transatlantic submarine cable completed in 1866, it still took time to transmit messages from the Old World to the New World. Thanks to the cable, the time for a message to cross the Atlantic decreased from ten days (by steamship) to a matter of minutes. For Kipling, messages passing along the ocean floor in the darkness represented the beginning of a new era, one in which continents are united, albeit in a gloomy Victorian age confronted with another result of Thomson's incessant calculations: the death of the sun.

Drawing on the first law of thermodynamics, combined with a conjecture concerning the source of the sun's energy and an estimation of its radiation rate, Thomson derived some disturbing new results. He announced that the sun had radiated energy for less than 100 million years, and that its remaining life was no greater than a few million years. He also made an independent estimate of the age of the earth's mantle. Using Fourier's heat theory and temperature gradients measured in mines and boreholes, he calculated that the formation of the mantle from its original molten state required a cooling period of 98 million years. Only a few years before Thomson communicated these results, Charles Darwin published *On the Origin of Species* (1859), in which he estimated, rather hastily, that the age of the earth was greater than 300 million years. Thomson's result thus appeared to rule out Darwin's theory on physical grounds, thereby inducing many readers to join Thomson in discounting Darwinian evolution. The death of the sun, while not imminent according to Thomson's calculation, was nonetheless a depressing prospect for the cultural mavens of the Victorian Age. This theme resonated in poems by Matthew Arnold, Rudyard Kipling, and others, though other more hopeful images also emerged. The popular success of Darwin's subsequent book, *The Formation of Vegetable Mould* (1881) has been linked by the literary critic Gillian Beer to the power of Darwin's imagery of the nearly blind worm, toiling in subterranean pathways. This served, in the minds of some Victorian readers, as an antidote of sorts to Thomson's troubling prediction of the death of the sun.[2] Kipling's blind white sea-snakes and message-carrying telegraph cables exercised a similar effect on contemporary readers, countering anxieties brought on by the results of Thomson's calculation of the sun's dwindling energy.

By the 1860s, the commercial importance of submarine cable telegraphy for Britain's empire helped resuscitate Faraday's field theory, which was no longer viewed as an arcane and fanciful model. As the century grew older, field theory and its inventors came to be revered as national treasures. In a somewhat more subtle fashion, the power of science and mathematics as tools for solving real-world problems gained increasing recognition among engineers, and not only telegraph engineers. Technical societies and engineering schools flourished in the latter half of the nineteenth century, in what became known as the Second Industrial Revolution. The spread of wired telegraphy from the 1850s, and the development of related electrical measurement devices, underlined the utility of standardized units of measurement.

In response to the spread of electrical instruments and machinery, in the early 1860s Thomson called for the formation of a BAAS committee on standards of electrical resistance. The new standards were to respect the principle of "absolute" units proposed decades earlier by Gauss and Wilhelm Weber. This meant that the definition of such units had to be independent of forces and fields that vary from place to place, such as the attraction of gravitation, or the intensity of the earth's magnetic field. The committee succeeded in proposing electrical units, although these were not widely followed at first. Nonetheless, some elements of the committee's recommendations were adopted by the first international electrical congress held in Paris in 1881, such as the nomenclature of "ohms" and "volts" alongside the ampere for current and the coulomb for charge (a current of 1 ampere per second produces 1 coulomb of charge). Maxwell was an active member of this commission, having become head of Cambridge's Cavendish Laboratory.

By the mid-1850s, Maxwell had become a leading exponent of Thomson's approach to theorizing via "physical analogies." Thus, he applied Fourier's heat theory to electrostatics and used Green's potential function to characterize Faraday's theory of dielectrics, among other powerful analogies. The young Maxwell found physical analogy well suited for realizing progress in theoretical physics. He described his method as a middle ground of sorts between a purely mathematical approach, on the one hand, and the formulation of physical hypotheses, on the other. This led him to embrace several fruitful analogies, including that between light undergoing refraction and a charged particle moving in a strong field of force, and a three-way analogy between light, the vibrations of an elastic medium, and electricity.

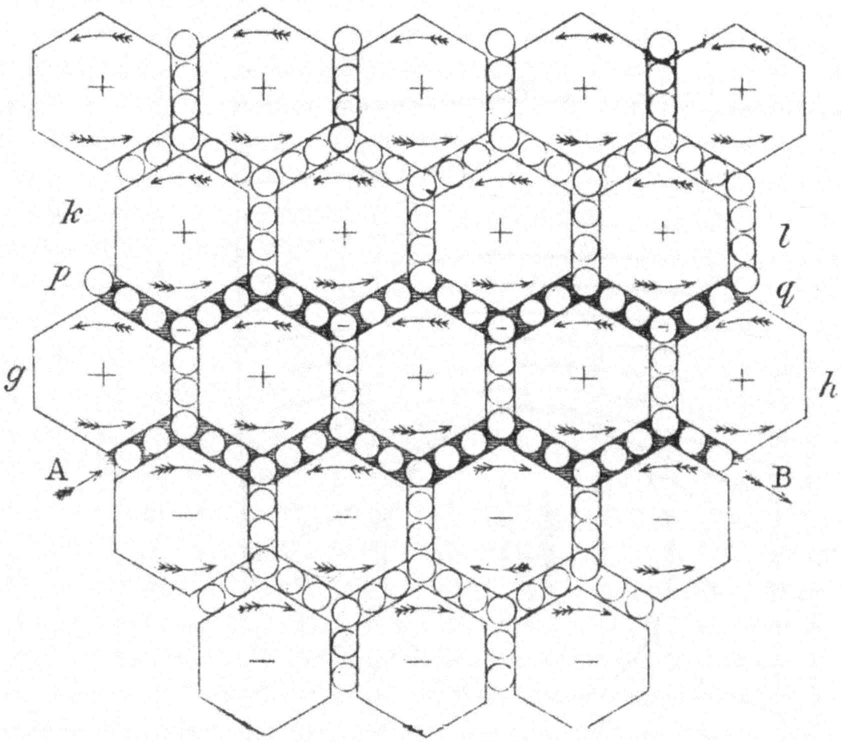

FIGURE 5.6. Maxwell's vortex-cell model of electric current (Maxwell, Scientific Papers, Vol. 1, 1890). Image courtesy of Scott A. Walter.

The physical analogy that Maxwell focused upon in his groundbreaking paper of 1855, "On Faraday's lines of force," was one between a system of electrical and magnetic poles with inverse-square interaction, and a field of incompressible fluid flowing in tubes, as suggested by Faraday's lines of force. Whereas Thomson had exercised prudence in his exploration of physical analogy, Maxwell pursued these with vigor, as this paper clearly reveals. Faraday, in his experiments in electrochemistry and on dielectrics, sought to measure electrical force along and across lines in terms of "intensity" and "quantity." Intensity referred to the tension that causes electric current, whereas quantity referred to the current strength and the integral current produced by a charged condenser. These concepts were central to Faraday's account of force and flux, which were the basis of his theory of electrification. In Maxwell's mind, quantity referred to the number of tubes of force intersecting a surface, while intensity referred to the number of surfaces intersected by a given tube of force.

Maxwell went on to develop a mechanical model of the electromagnetic field, inspired in part by Thomson's image of space-filling molecular vortices. Alignment of such vortices, Thomson suggested, resulted in magnetization of matter. Maxwell's mechanical model likewise utilized vortices to illustrate both magnetism and electric current, the latter being generated by a "layer of particles, acting as idle wheels." These rolled without sliding between the cellular vortices of the elastic, space-filling medium. The resulting system of field equations, equivalent to what would later be known as "Maxwell's equations," unified electrostatics and electrodynamics, but that was not all. Following Thomson's lead, Maxwell considered transverse waves in the medium, which he assumed to be composed of spherical cells whose elasticity arises from intermolecular forces. The velocity of these transverse waves turned out to be the velocity of light (Siegel, in Cantor & Hodge 1981).

While Maxwell's argument in favor of a common foundation of optics and electrodynamics later turned out to be faulty, in the meantime he shelved his magneto-mechanical model in favor of a Lagrangian foundation for his field equations, thereby realizing the unification suggested by the earlier model. Maxwell's ambitious theory inspired a number of British physicists to explore its consequences, both physical and mathematical, although Thomson considered this a step backward. Over the years, these "Maxwellians," as they called themselves, convinced many of their compatriots to forgo the older notions of electric fluids and action-at-a-distance in favor of field-theoretic concepts. On the Continent, however, where alternative approaches held sway, Maxwell's theory sparked little interest. Helmholtz was a notable exception, and in 1879 he

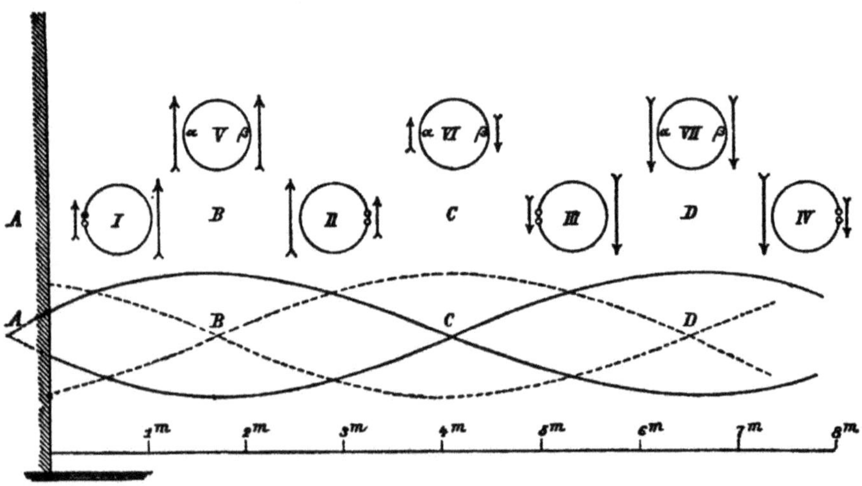

FIGURE 5.7. Hertz's measurement of a standing wave in air (Hertz, Electric Waves, 1893). Image courtesy of Scott A. Walter.

encouraged his student Heinrich Hertz to engage with British field theory. A brilliant experimenter and theorist, Hertz recognized that a source of high-frequency electric oscillations could provide decisive evidence supporting the validity of Maxwell's theory. While preparing his lectures on electricity and magnetism at Karlsruhe Polytechnic, Hertz noticed a spark discharge phenomenon, which suggested the existence of rapid electric oscillations. Hertz quickly mastered a device for producing such oscillations in a controlled manner, and used it first to generate electromagnetic waves on wires, and then, in 1887, electromagnetic waves in the air.

Hertz's experimental setup was not difficult to replicate, so within a short time physicists across Europe confirmed the existence of electromagnetic waves. Hertz's discovery came as no great surprise to Maxwellians, but others found it quite shocking. Almost overnight, it launched a new field of research, as physicists, engineers, and mathematicians began investigating phenomena associated with Hertzian waves. Within a few years, they also began contemplating military and commercial exploitation of wireless waves in air, as Guglielmo Marconi and others began filing the first patents on wireless telegraphy. By 1902, telegraph messages were not only snaking their way across the ocean floor in submarine cables, they were flying through the air—and around the world—on Hertzian waves.

CHAPTER SIX

Mathematics and Technological Change

Connecting the World via Calculation, Observation, Transport, and Communication

JUNE BARROW-GREEN AND TONY ROYLE

Mathematics underpinned much of the technological change witnessed throughout the nineteenth and early twentieth centuries. Advancements in transportation and innovative applications of electromagnetism provided a great deal of the impetus that would revolutionize many aspects of society. These two predominant threads within the fabric of technology permeated and deeply influenced other areas of technological progress. The ability to move information, goods, and people over ever-increasing distances in a timely fashion fostered efficient communication of ideas, bolstered economies, and enabled more frequent gatherings of scientific communities. Emergent from these exchanges and collaborations were the nascent propositions and designs that would provide the stimuli and ingredients for transformation. Some of the associated mathematics was new, but much that already existed found novel application. The improving precision of mathematical instruments also played a key role in progress by enabling the accumulation of more accurate and abundant data and by allowing observations that offered insight to inform theory and design. Mathematicians themselves were prime movers in the transformation, many diversifying or adapting to become leaders in the various strands of engineering and physics that lay at the heart of this flux. Some would become household names whilst others remained more anonymous, but whatever their profile the hallmark of their influence is stamped on almost every facet of the innovation that drove technological advancement as the nineteenth century unfolded.

TRANSPORT AS A CATALYST FOR CHANGE

Few would argue that the static steam engine was the salient device that powered the Industrial Revolution of the early 1800s, but it was the engine's application to transport that sparked a fundamental change in methods of communication. The ability to move people and goods more efficiently over land and sea had a remarkable impact on life in general and radically influenced the way information was disseminated and exchanged. In 1800, for example, it took 60 hours by stagecoach to travel from London to Edinburgh. but by the end of the century the railway had reduced this time to a mere nine hours.

James Watt was the man responsible for improving the efficiency of the steam engine beyond that of the original designs of Thomas Newcomen that dated back to the early 1700s (Woolmar 2008). Watt's epiphany was the realization that a separate condenser would reduce the loss of latent heat from the system, thereby helping to preserve the amount of steam available during each piston cycle. Watt was also fascinated by the theory of linkages, the means by which one type of motion could be converted to another, and he famously devised a system for converting the rocker beam movement in a pumping engine into the rectilinear motion required for a piston shaft. Russian mathematician Pafnuty Chebyshev also found the theory of linkages intriguing and set out to find new mathematical methods that would define the connections necessary to produce, with great precision, the desired type of motion. Chebyshev's mathematical approach to solving linkage problems resided in the field of polynomial approximation of functions, a domain of mathematics now synonymous with his name. Improved and more versatile linkages played a significant role in facilitating development of the steam locomotive throughout the nineteenth century.

The Stephenson-built *Locomotion No. 1* began hauling passengers on the Stockton and Darlington Railway in 1825, an enterprise that soon expanded to provide inter-city services across Britain. The opening of the first line between Liverpool and Manchester in 1830 was somewhat overshadowed by the unfortunate death of the Member of Parliament for Liverpool, William Huskisson, who was struck by Stephenson's *Rocket* at an intermediate water stop between the two cities. Huskisson's demise, however, was reported around the globe in the popular press, inadvertently announcing the arrival of the steam train as a primary mode of mass transportation. Within weeks of the inaugural services, the potential for expansion was being exploited. By 1850, over 6,000 miles of track had been laid in Britain, creating a network joining all the major centers of population and providing an impetus and inspiration for related developments on the Continent. Franz Josef von Gerstner, a professor of mathematics at the Prague Polytechnic, who built the first steam engine in Czech lands, proposed building a railway between Budweis (today, České Budějovice) and Linz.

In many ways, George Stephenson and his son Robert epitomized the genre of British engineers who were responsible for driving technological progress in the early 1800s. George had received nothing more than a basic school education, a disadvantage that isolated him from academic circles in London and Cambridge, prompting him to ensure that his son would fare better. Robert thus attended the Percy Street Academy in Newcastle before joining the *Newcastle Literary and Philosophical Society*. The latter allowed both father and son access to the technical books that they would use to inform the construction of their steam engines and railway networks. This was a very different form of mathematical and technical education than that which heavyweight mathematicians, engineers, and scientists would obtain later in the century. Indeed, the role of the many Literary and Philosophical/Scientific Societies in Britain's cities and towns was a central one in providing access to information and forums for debate and discussion in the era preceding the dominance of university-based influence.

The engineers who first dealt with the mechanical applications of steam power, however, were not overly concerned about the fundamental properties of the heat responsible for producing mechanical energy. Yet, without such understanding, it became impossible to analyze the performance and efficiency of these machines. In 1824, Nicolas Sadi Carnot used the caloric theory of heat of Antoine Lavoisier to study the efficiency of steam engines, out of which the concept of a reversible process was born. From Carnot's work fell the seeds that would subsequently grow into the second law of thermodynamics and the concept of entropy, with all the associated science (see Chapter 5). Rudolf Clausius, along with William Thomson (later Lord Kelvin), William Rankine, and Hermann von Helmholtz pioneered this terrain in the early 1850s. They developed mathematical equations to describe dynamical theories of heat related to conservation of energy, expressions that eventually led to the first law of thermodynamics. James Clerk Maxwell in the 1860s, and later Ludwig Boltzmann, both worked to refine the elegance and extent of the mathematics, paving the way for diverse and extensive applications in engineering and science.

One of the early challenges that interconnectivity across vast distances presented concerned the accurate surveying of the terrain over which the railways travelled. This brought to the fore the mathematics associated with geodesy, particularly levelling (the process of determining vertical levels relating to Mean Sea Level to assist with accurate construction and cartography). The

First Geodetic Levelling of England and Wales took place in parallel with the expansion of the railways, and similar initiatives sprang up in many other countries. In France, the École Polytechnique (EP) was the source of the necessary mathematical expertise. At the EP, large numbers of students in mathematics, physics and engineering were taught by some of the country's foremost mathematicians. It was here that many military and civil engineers began their training before proceeding to one of the several specialist Écoles, such as the École des Mines and the École Nationale des Ponts et Chausées. Graduates of the EP include Charles Lallemand who ran the French levelling project, and the engineer Charles Maurice d'Ocagne, the inventor of the nomograph, a graphical calculating device developed originally to expedite the complex cut-and-fill calculations for earth removal required for the construction of the French railway system.

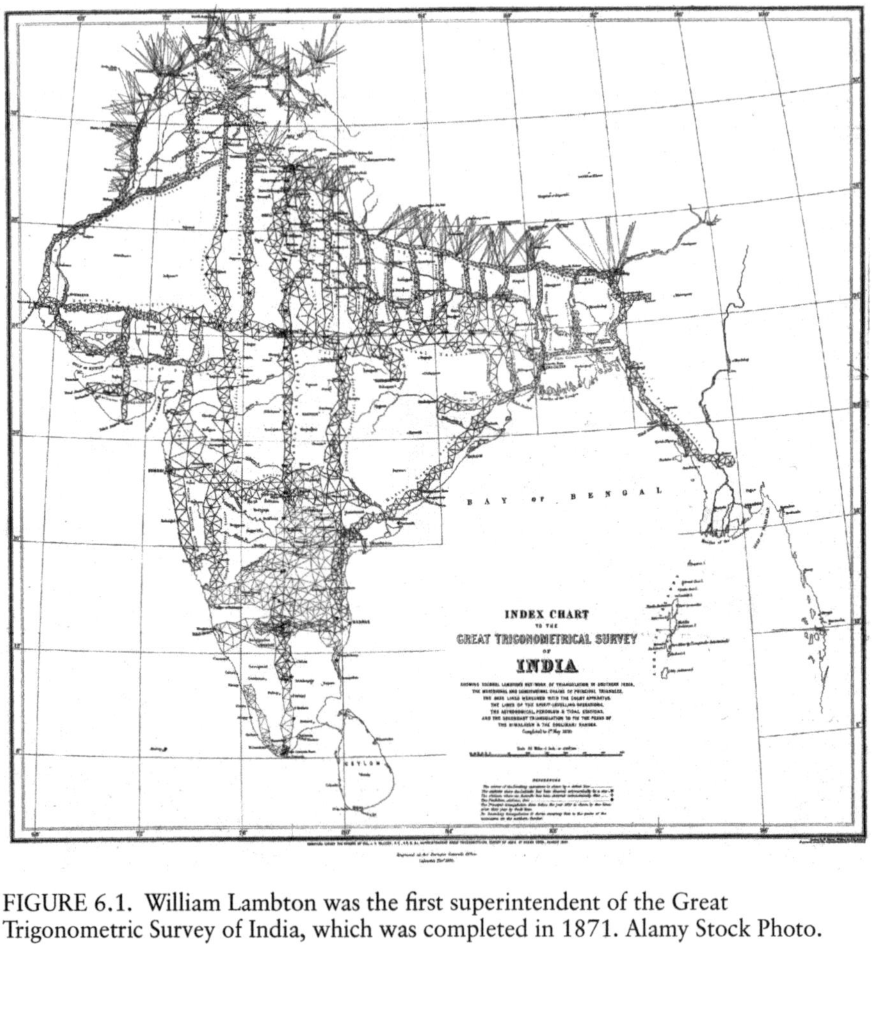

FIGURE 6.1. William Lambton was the first superintendent of the Great Trigonometric Survey of India, which was completed in 1871. Alamy Stock Photo.

In India, William Lambton commenced the Great Trigonometric Survey. Lambton, had been taught mathematics by Charles Hutton, and put this schooling to good use during his attempt to survey the entire Indian subcontinent. Hutton, who taught Lambton at Northallerton Grammar School in Yorkshire, subsequently became professor of mathematics at the Royal Military Academy (RMA) at Woolwich in 1773, and is perhaps best remembered for his attempts to calculate of the density of the earth. Lambton was succeeded as superintendent of the Survey by George Everest, who had also been educated at the RMA. In 1865, the Royal Geographical Society named the world's highest mountain peak in Everest's honor. Everest was followed by Andrew Scott Waugh and then by James Walker, both of whom learnt their mathematics at Addiscombe Military Seminary from Jonathan Cape. Cape was a teacher notable for being up to date and whose academic course, based on that of Hutton, included many recent developments and innovations from France. Waugh was recognized for his mathematical skills while Walker was responsible, in 1871, for the first completion of the Survey.

Investment in transport infrastructure was endemic across the world during the nineteenth century, much of the mathematical influence and inspiration originating in France. In 1795, Gaspard Monge in the introduction to his lectures on descriptive geometry at the EP had emphasized the importance of precision for the French engineer, although his aspirations for the use of higher mathematics in industrial application remained largely unfulfilled at the time. This type of geometry did provide, however, the stimulus for engineer and mathematician Jean-Victor Poncelet to develop and pursue his work in projective geometry, a subject and mathematical tool that would become an intrinsic element of the rather mathematically sophisticated engineering education on the Continent. It led in the longer term to methods such as graphical statics, founded by Carl Culmann, who cut his teeth at the military engineering school at Metz before enrolling at the EP. Culmann, who had been inspired by the graphical methods of Poncelet, made use of improving transportation systems to make a tour of the United Kingdom and the United States so that he could conduct a comparative study of truss bridges to inform his book, *Die Graphische Statik*. The work found application in fields such as crystallography and civil engineering, the latter exemplified by the construction of the Eiffel Tower in 1889. Meanwhile in England, the German mathematician Olaus Henrici, who had settled in London, combined his mathematical education with his experience in an engineering business to produce *Skeleton Structures, Especially in Application to Building Steel and Iron Bridges*. Henrici's book later found an audience amongst some of the engineers who built skyscrapers in New York City.

The steam train gradually replaced the horse and carriage as the most expeditious means of transport, although it would be decades before the rail network matched that of the roads in terms of geographical coverage. Alongside steam-driven trains, automobiles employing the same means of propulsion were developed, but the impracticality of their use restrained proliferation and impact. This mode of transportation was waiting for an efficient version of the internal combustion engine to appear and—unique to Britain—the repeal of the Locomotive Act of 1865, which required self-propelled vehicles on public roads to be preceded by a man waving a red flag and blowing a horn to alert unsuspecting pedestrians!

The mathematics behind the steam revolution was predominantly that needed to calculate and describe the forces experienced by the various components and moving parts that comprised the vehicles, engines, and associated infrastructure. Stresses and strains placed upon the girders of the new bridges being built to support railway tracks came under mathematical scrutiny, as did those felt by the pistons and gear assemblies that turned various crankshafts and wheels. In the 1820s, Claude-Louis Navier developed the theory of elasticity and made its descriptive mathematics practical to use. He also better defined the role of the structural engineer as someone who should employ mathematics and materials science to ensure the integrity of structures during the design and construction phases of a project. Later in the century, Carlo Alberto Castigliano introduced his least strain energy methods to aid the computation of the displacement of structures under load using partial derivatives of strain energy, a technique that would subsequently find application in aeronautics and many other genres of engineering.

The advent and spread of the inter-city rail network in Britain coincided with the recognition that action was needed to re-establish the country's status in science to bring it on a par with its European neighbors. Such was the malaise that the Lucasian Professor at Cambridge, Charles Babbage, was driven to publish his *Reflections on the Decline of Science in England* (1830) (see Chapter 4 for Babbage's earlier role in promoting the Cambridge "Analytical Society"). Babbage gave a damning indictment of the country's ability to prosecute meaningful science, particularly in fields with mathematics at their base, the target of his ire being the apparently moribund Royal Society. The consequent founding of the British Association for the Advancement of Science (BAAS) in 1831 was thus welcome news to the greater scientific community in Britain, and it soon provided a forum that would reap the benefits of relatively easy opportunities for communication and travel for its members. Annual meetings were held, each in a different city, bringing the latest science and mathematics to a larger public. By the end of the century the BAAS was venturing into the reaches of Empire, holding meetings in Montreal (1884) and Toronto (1897).

Whilst the ability to move across land was experiencing change beyond recognition, revolution in the air had yet to materialize. Practical use of powered aircraft as a means of transportation was a century away, but as early as the first decade of the nineteenth century, engineers and mathematicians were already laying the foundations. These challenges required a new breed of mathematician versed in aeronautics, and there is reasonable justification to consider George Cayley as the first Englishman who could legitimately be categorized as an aeronautical engineer. His pedigree as a mathematician per se is difficult to establish, but his schooling by theologian and mathematician George Walker, and engineering competence by example, indicate that he had a firm grasp of applied mathematics. The aeronautical exploits of Cayley, who was a distant cousin of British mathematician Arthur Cayley, dated back to the final decade of the eighteenth century. His first aeronautically themed publication of note was his three-part text, *On Aerial Navigation*, which was serialized in *Nicholson's Journal of Natural Philosophy, Chemistry, and the Arts* (1809–10). Cayley constructed his first full-size glider in 1849, initially testing it using ballast, and then with a servant's 10-year-old son aboard. His attempt in 1853 to launch his footman, John Appleby, into the air using a small glider (a modified tri-plane variant) being towed down a small hill by a posse of volunteers is recognized as a significant milestone in the record of manned flight. Cayley's footman flew some 900 feet across Brompton Dale, near Scarborough, Yorkshire. There is some debate surrounding the identity of the "pilot" that day, but there is no doubt that the flight took place, details of the so-called "Governable Parachute" which was employed having appeared a few months earlier on the front page of the popular periodical *Mechanics' Magazine*.

Just as on land, travel across water became safer and more efficient as mathematics began to influence vessel construction, propulsion, navigation, and the ability of those on board at sea to communicate across vast distances. Mathematics also had an impact on the understanding of the tides and accuracy of tidal measurements and associated data collection. The ability to predict the tides proved important for numerous reasons, not least to ensure the safe and accurate navigation by shipping of ports and shallow coastal waters. In the United States the value of mathematical know-how in this regard was not lost on the hierarchy of the US Navy, which established a Corps of Professors of Mathematics in 1848 to instruct staff at the US Naval Academy, the Naval Observatory, and indeed aboard ships at sea. In Britain, William Froude, a graduate in mathematics from the University of Oxford, became the doyen of testing to define optimum design for ships' hulls. Froude's grasp of hydrodynamics allowed him to construct and test scale models and derive suitable scaling factors to guide the construction of full-size ships, and in 1872 he oversaw the commissioning of the Admiralty's first naval test tank in Torquay. One of Froude's later inventions was the water brake dynamometer that was used by the British Admiralty to determine the power output efficiency of the marine engines used by the Royal Navy.

Also closely connected with navigation was the basic gyroscope invented in the 1760s by the Tübingen professor of mathematics and astronomy, Johann

Bohnenberger. This prototype was further developed by Pierre-Simon Laplace and Leon Foucault in France and was eventually combined with the electric motor to become the heart of the gyrocompass. This new means of holding accurate headings in craft at sea or in the air was later fully exploited by the American inventor, Elmer Sperry. As an expert in dynamo technology, Sperry attempted to integrate the power of electricity with various systems and equipment used in both mining and transport. He developed electric mining equipment to improve the efficiency of deep coal extraction, electric systems for the railways, and even an electric car that ran on lead-acid batteries. A spectacular demonstration of the potential use of the gyroscope involved Sperry's son, Lawrence. In 1914, he piloted a single-engine Curtiss C-2 biplane equipped with gyroscopically stabilized flying controls down the course of the River Seine in Paris in front of a huge crowd. On his first pass he removed his hands from the controls to demonstrate the control automation. Then, on the second pass his passenger, the French mechanic Emil Cachin, stood out on the starboard wing to show how the gyro compensated for lateral imbalance. For the finale, on the third pass Sperry himself got out of the cockpit and stood on the port wing, leaving those watching in no doubt as to the capability of the rudimentary autopilot.

On land as well as sea, huge changes took place in the nineteenth century that promoted opportunities but also the propensity to travel. Powered initially by steam, but latterly by the internal combustion engine and the electric motor, various forms of transport evolved connecting people who had previously remained unconnected. The efficient movement of goods had a major impact on trade and economies. Travel by air, however, had yet to be fully exploited, something that would rapidly change at the dawn of the twentieth century.

INSTRUMENTS FOR COMPUTATION, ANALYSIS, AND OBSERVATION

Charles Babbage was one of several observers who pointed to increased competition between nations in an age of expanding industrialization. In his book entitled *On The Economy of Machinery and Manufacture*—which originated from over a decade spent in research for his first automated calculating machine, Difference Engine No. 1—he stressed the necessity of the "science of *calculation*." What Babbage sought in constructing his Difference Engine was a fully automated way of producing reliable mathematical tables: a machine that would both do the calculations and print the results, thereby eliminating the potential sources of error (Swade 2000).

Whilst the mathematical concepts and engineering ingenuity required to develop such automation were in place by the 1830s, the significant funding needed to back such a project was less forthcoming. This proved to be the

FIGURE 6.2. Babbage's Difference Engine Number 2 with Printing Mechanism. This engine was constructed by staff at the Science Museum in London based on Babbage's designs drawn up circa 1848. Photo by SSPL / Getty Images.

stumbling block that prevented Babbage's project from reaching fruition. It was at this time that mathematician Ada Lovelace began her pioneering contributions to the science of computing (Hollings, Martin, and Rice 2018). Introduced to Babbage by her private tutor, the polymath Mary Somerville, Lovelace took a keen Interest in the workings of Babbage's second and more ambitious project, his Analytical Engine, a general-purpose programmable machine, though it, too, was never built. This led to her renowned article on the machine, a translation of a paper in French by Luigi Menabrea to which she added an extensive commentary giving far-sighted predictions about the potential capabilities of the machine.

The mechanization of calculation was not an exclusive idea to Britain, however; the first commercially successful mechanical calculating device was produced by Frenchman Thomas de Colmar, who identified applications of such a machine in simplifying the business of insurance. His bespoke *Arithmomètre*, patented in 1820, was based on the stepped drum design introduced by Gottfried Wilhelm Leibniz. This device was produced for the general market from 1851 onwards, eventually bringing great wealth to its inventor. A major step forward in this type of mathematical instrument came towards the end of the nineteenth century when Swiss engineer Otto Steiger patented a device capable of direct multiplication, the *Millionaire* calculator. Earlier machines were based on additive methods to solve multiplicative problems, and so were much slower in producing results. Thousands of these instruments were exported across the world, finding homes in the realms of business, science, government, and the military. Not only did the improved speed of computation influence technological change, the level of accuracy achievable over realistic time frames played its part in making once intractable calculations not only feasible but practical as well.

Related to these activities was the need for accurate estimates of planar areas, a problem facilitated by the progressive development of planimeters throughout the 1800s. These instruments, which estimate the area of an enclosed planar region by tracing its boundary, were used extensively by engineers and surveyors. Aside from their utility for scientific drawing, they were also used for determining the area of an indicator diagram—the chart used to estimate the power of a steam engine—and also for calculating areas in land surveys. Although first conceived in the eighteenth century, the first such device, also called a mechanical integrator, made its appearance in Bavaria in 1814. Over the course of the century, these instruments inspired the likes of William Thomson and Maxwell to exploit and improve their calculating power. One such example, designed by the Scottish civil engineer John Sang, "greatly excited" the young Maxwell when he saw it on display at the Great Exhibition in 1851. Convinced there were ways to improve it, he developed one of his own. The principles behind it were published in 1855, thereby paving the way for other new types of planimeter. One of the most popular, due to its simplicity and affordability, was the polar planimeter invented in 1854 by the Swiss mathematician and instrument-maker Jakob Amsler. Amsler so named it because it turns around a fixed point, i.e., the pole, of the instrument. In 1857, Amsler opened a workshop in Schaffhausen, eventually selling more than 50,000 polar planimeters of six different designs to customers across the globe during the course of the next 60 years. Indeed the harmonic and differential analyzers that became prevalent in the early 1900s had modified planimeters as their central components. Henrici, in his 1894 BAAS *Report on Planimeters,* stressed the importance of machines capable of mechanical integration.

As already mentioned, an important metric pertaining to shipping activities was knowledge of the tides. During the 1870s, Thomson produced a series of increasingly more efficient tide-predicting machines that used harmonic data to predict future tides. These worked by repeatedly evaluating a sum of upward of 20 trigonometric terms, and doing so for different values of time, and translating these calculations into a continuous graph of the height of the tide as a function of time for a given location. The tide predictor marks one of the most successful

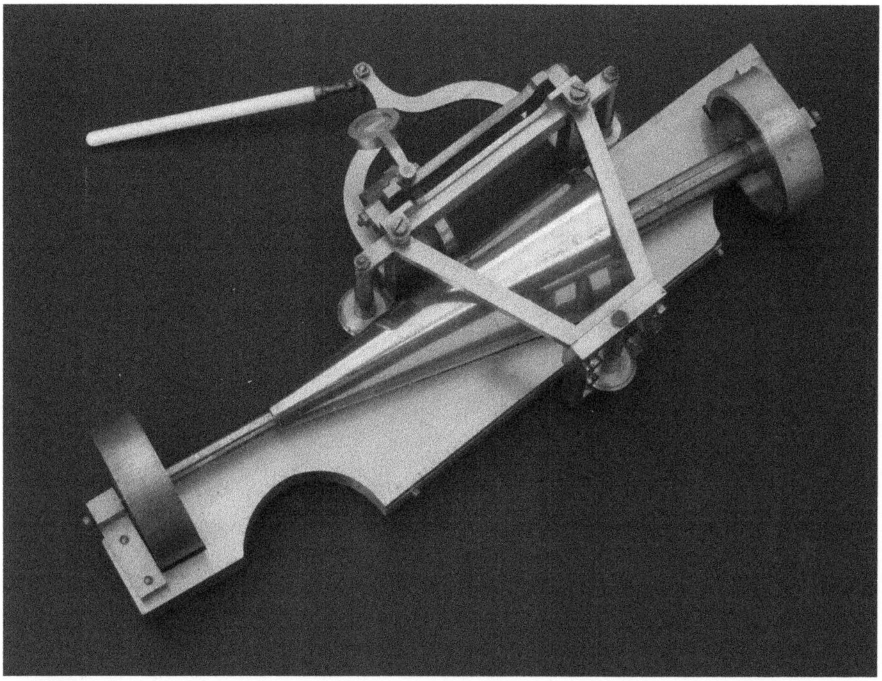

FIGURE 6.3. John Sang's planimeter, which received an honorable mention at the Great Exhibition of 1851. Photo by Science & Society Picture Library / SSPL / Getty Images.

introductions of automatic calculations into mathematics, remaining operational well into the twentieth century. It was used, for example, to calculate the tides for the Normandy landings on D-Day in 1944.

In the same period, William Thomson also produced a tidal harmonic analyzer adapted from the mechanical integrator designed by his elder brother James. Its objective was to "substitute brass for brain" in the great labor of calculating the elementary constituents of the tidal rise and fall; the mechanical integrator was based on a simplified version of Maxwell's planimeter. With the harmonic analyzer, William Thomson solved the problem of extracting the harmonic components from a tidal curve; something first demonstrated in 1876. In 1878 the same instrument, but on a larger scale, was brought into use at the Meteorological Office. Another designer of harmonic analyzers was Henrici. His version of the device, which was smaller and thus easier to transport than Thomson's, was endorsed by Felix Klein, and later used in the field of acoustics. Despite its cost (equivalent to about £7,000 in 2021), several academic institutions bought this state-of-the-art instrument, including the École Polytechnique in Paris, the Tomsk Institute of Technology in Russia, and the Tokyo Earthquake Investigation Committee in Japan.

FIGURE 6.4. Olaus Henrici's Coradi Type III Harmonic Analyzer, 1894. Photo by SSPL / Getty Images.

Mathematics was intertwined with astronomy throughout the century in its predictive and descriptive capacities, and took a central role in the functioning of the astronomical observatories that were appearing across the globe (North 2008). William Parsons, 3rd Earl of Rosse, was an astronomer who brought an education founded in mathematics to bear in conducting astronomical work. Following his studies in Dublin and Oxford, he had built in Ireland during the 1840s, what was then, the largest telescope in the world that he used to study, amongst other things, spiral galaxies.

The difference this mechanization of mathematics made to life in the late nineteenth century can be compared with the way computing power has influenced technology in the modern era; data gathering, handling, and processing changed beyond recognition. Calculations that had previously been impractical became routine. What is more, these new capabilities carried significant economic and socioeconomic implications. Just as computers and automation affect the nature and number of jobs today, so this initial tranche of mathematical instruments presented similar challenges and opportunities to contemporary society. Mathematics through technology was an instigator of change, and an essential tool of progress.

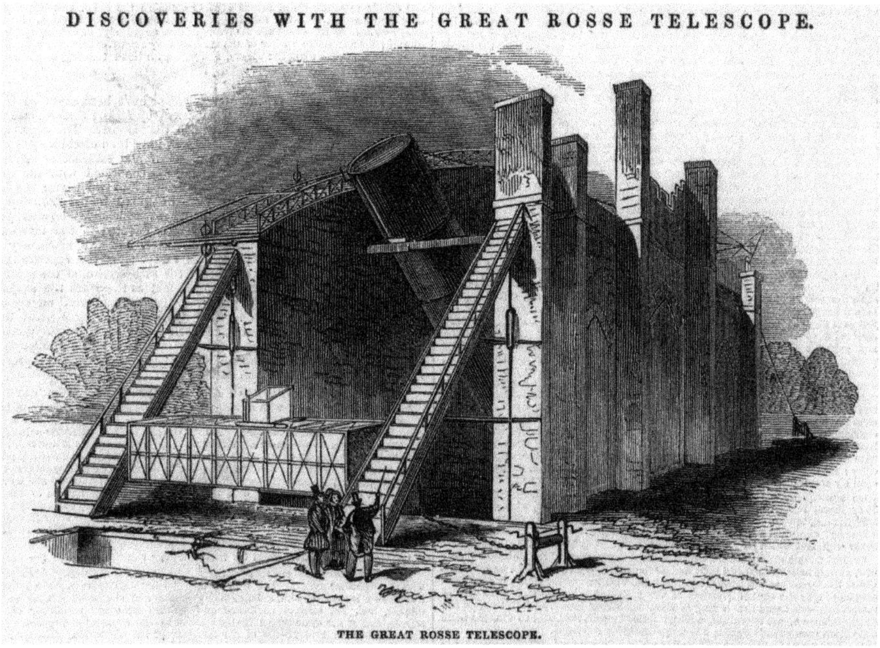

FIGURE 6.5. Engraving from the *Illustrated London News*, showing the Great Rosse Telescope built during the period 1842–5 by William Parsons, 3rd Earl of Rosse. Photo by SSPL / Getty Images.

EDUCATION AND THE TRANSMISSION OF IDEAS

At Cambridge, the center of British mathematical education in the nineteenth century, large numbers of young men were intensively trained in mathematics as a basis for their careers (technical and otherwise). The Cambridge Mathematical Tripos (MT) was the most prestigious examination in Britain and competition to succeed was fierce. To be a leading wrangler, as a student in the first class was called, opened the door to a career of choice, be it in academia, the law, the church or elsewhere. Several of those who were particularly successful, such as George Stokes and the aforementioned William Thomson and James Clerk Maxwell, would come to dominate British mathematical physics in the second half of the nineteenth century, and bring mathematics to bear on significant elements of technological change, such as telegraphy and electricity. The MT was also a springboard for mathematicians going to the colonies and establishing mathematical outposts and observatories. By the end of the century, the expansion of engineering had led to a broadening of subjects taught at Cambridge and with it, in 1893, the establishment of the Mechanical Sciences Tripos, a natural evolution that helped to put applied mathematics on the academic map.

FIGURE 6.6. A Clarke, Chapman, Parsons & Co Steam Turbine Generator, 1884. Photo by SSPL / Getty Images.

The benefits of such a mathematical training and its impact on technological development is evident in the case of the inventor Charles Parsons, son of William. Educated at Trinity College, Dublin, and an 11th wrangler, Charles, rather unusually for someone with his family pedigree, went straight from university into industry. There, in 1894, he designed and manufactured the first of his steam turbines. By linking the turbines to electrical generators these machines could not only be used to provide motive force, but also to generate cheap and abundant electricity. The availability and affordability of Parsons' turbines served to revolutionize the design and capacity of marine vessels, and they soon functioned as the hearts of iconic ships such as the RMS Mauretania, the largest liner afloat in 1906, and her elder sister, the RMS Lusitania. Such liners could hold speeds of 25 knots on the ocean, although even this was at times insufficient. On May 7, 1915, the Lusitania was unable to evade the torpedo fired from a German U-boat that sank the liner. Three years earlier, an even more famous sinking of a ship partially powered by a Parsons' steam turbine took place: on April 14, 1912, the RMS Titanic went down on its maiden voyage after striking an iceberg south of Newfoundland. Other rather more menacing ships to benefit from Parsons' invention were the Dreadnought battleships of the British Navy.

The impact of large steam-powered turbo generators in providing cheap and reliable electricity to industry, households, and to illuminate towns and cities across the world, can hardly be overstated. The Newcastle and District Electric Lighting Company, set up in 1889 in the northeast of England, paved the way for global proliferation. This new generation of steam turbines began to replace the more cumbersome, noisy, and inefficient reciprocating engines utilized in the first tranche of central generating plants, such as the one set up by Thomas Edison in New York's financial district. Parsons' designs had the huge benefit of producing rotating motion directly to turn the armatures of any attached electrical generator, whereas the reciprocating engines required some linkage to convert linear motion into rotational. In addition to the clear advantages in efficiency, therefore, the potential to scale up the steam turbines to offer greater electrical output at modest cost made them very attractive entrepreneurial propositions (Scaife 2000).

Another influential protagonist in the field of marine engineering was John Thornycroft, who benefitted from study at Glasgow University under Thomson and Rankine. He went on to become a pioneer in the design and development of torpedo boats and other fast-moving ships. The Cambridge-educated Charles Rolls equally cast influence in the motor vehicle and aeronautics industries around the turn of the century. Partner in the famous Rolls-Royce concern that was founded in 1904, Rolls would attract infamy as the first Briton to die in a crash involving a powered aircraft when the tail of his Wright Flyer broke off during an aerial exhibition near Bournemouth.

The Cambridge wrangler and polymath, Henry Fox Talbot, played a seminal part in the development of photography and spectral analysis. He wrote papers on elliptical integrals prior to finding an interest in the study of light through a friendship with John Herschel, Senior Wrangler in 1813 and son of famous astronomer William Herschel. John Herschel augmented Talbot's achievements and is credited with the first use of the word "photography." The ability to record images was a major addition to the armory of nineteenth-century technologists.

Elsewhere in Europe, from the middle of the century, the Technische Hochschulen (TH), notably those in Karlsruhe, Munich, Prague, Vienna, and Zurich, promoted the mathematization of the engineering sciences. At Karlsruhe, Ferdinand Redtenbacher brought mathematics into the teaching of engineering. One of his students was Franz Reuleaux, who inaugurated the study of kinematical models at the TH Berlin. Others at Karlsruhe included Christian Wiener and Alfred Clebsch, who taught Henrici. At Harvard, Benjamin Peirce taught mechanics as professor of mathematics from 1831 until his death. From 1867 to 1874, he was also Director of the US Coast Survey, an institution which had a very wide remit during the nineteenth century, being for much of that time the only publicly funded scientific institution.

The second half of the century also saw the development of laboratories for research and teaching in physics and engineering. In Britain, for example, the Clarendon Laboratory in Oxford opened its doors in 1872, since the Clarendon Trustees preferred spending their funds on a physics laboratory rather than on a riding school (the original destination of the bequest). Two years later, the Cavendish Laboratory at Cambridge opened with Maxwell as its first professor. In London at University College (UCL) the engineer Alexander Kennedy, another proponent of the importance of mathematics for engineering, pioneered the teaching of engineering using a laboratory (he also translated Reuleaux's book on kinematics). Later, Henrici—formerly of UCL and brother-in-law of Kennedy—opened a laboratory of mechanics for his students at the newly founded Central Technical College in South Kensington. In Edinburgh in 1913, Edmund Taylor Whittaker opened a mathematical laboratory where he pioneered the teaching of numerical analysis in Britain. These institutions provided facilities and environments that fostered progress in applied mathematics and engineering that fed into the technological progress experienced by society at large. They were sources of innovation and places where some of the best mathematicians were able to ply their trade and pass on their expertise.

Adding to the throughput of individuals versed in mathematics who would influence technology were scholastic institutions such as the Dockyard Schools in Britain. As early as 1806, maritime authorities recognized the need for individuals who had a solid grounding in mathematics applicable to ship design and operation. The first School of Naval Architecture, established in Portsmouth in 1811, thus began a tradition that would proliferate and endure, producing individuals educated in a way that provided an invaluable stream of bespoke expertise into the world of shipbuilding and general engineering.

The nineteenth century also saw the founding of many new national scientific and mathematical organizations, including the aforementioned BAAS, the American Association for the Advancement of Science, the London Mathematical Society, and the Société Mathématique de France, among others. Existing ones, such as the Royal Society and the Académie des Sciences, also persisted. Other more specialized societies, such as the Statistical Society of London (later the Royal Statistical Society), the American Statistical Society, and the Astronomical Society of London (later the Royal Astronomical Society) were also formed. They all provided venues for the exchange and display of ideas, through meetings and the publication of journals.

The period also witnessed the appearance of many independent scientific and mathematical journals although not all these survived. In addition to the educational changes that were taking place, this growth can be attributed to technological developments both in printing and paper production and the new distribution networks, especially the railway. An example is *The Analyst*, which was published in Des Moines, Iowa, from 1873. That *The Analyst* succeeded where other independent mathematical journals had failed can be attributed in part to the recent expansion of the telegraph and the railroad. This enabled many of the most eminent scientists and mathematicians from the East Coast to travel a distance of over a thousand miles to Des Moines for the 1869 Solar Eclipse and at the same time meet the man who would become *The Analyst*'s editor, Joel Hendricks. In this way Hendricks, a surveyor with a love of mathematics but no formal mathematical training, was able to establish a network of future contributors to his journal. In 1884 when Hendricks' health began to fail, he was able to arrange for the journal to move to the University of Virginia, where it was renamed the *Annals of Mathematics*. Under the editorship of Princeton's Solomon Lefschetz it later became one of the world's leading mathematics journals.

As the movement of people and objects across countries and continents became more practical, business interests began organizing gatherings that promoted technological awareness and stimulated progress. In Paris, the Exposition des Produits de l'Industrie Francaise was held eleven times between 1798 and 1849. One of the exhibits in 1798 was a mechanical device demonstrating the divisions of the metric system, which had been introduced by the Revolutionary government three years before (see Chapter 1 in Volume 4). A highlight in 1801 was the Jacquard loom, which was controlled by punch cards (Ada Lovelace would famously compare it to Babbage's Analytical Engine). At mid-century, Britain staged The Great Exhibition of the Works of Industry of All Nations (or The Great Exhibition) in London in 1851, an event organized by Henry Cole, a civil servant and inventor—today more famous for producing the world's first commercial Christmas card—and keenly promoted by Prince Albert, Queen Victoria's consort. Over 6 million people, more than a third of the population of Britain at the time, visited the Exhibition to see the 13,000 exhibits on display.

Public exposure to the mathematical sciences gained momentum during the last quarter of the century, which saw two important exhibitions for mathematics and related sciences. These were the 1876 loan exhibition of scientific apparatus in South Kensington, London, and the 1893 exhibition of mathematical models, apparatus and instruments in Munich organized by the German Mathematical Society. These events were open to the general public and attracted many people. The South Kensington exhibition, which was open for six months and had some 20,000 scientific artefacts on display, attracted over a quarter of a million visitors. Although much smaller with some 300 exhibits, the Munich exhibition was seen by about a thousand visitors during the month it was open. In that same year, mathematics was also in evidence at the Chicago World's Fair. There, visitors could attend a Congress showcasing mathematical research from the newly founded University of Chicago or visit the extensive display of

German mathematical models and literature on show, the latter arranged by Klein, Germany's official mathematical representative. Four years later in 1897, over 200 mathematicians gathered in Zurich for the first International Congress of Mathematicians, beginning a tradition which continues to this day. These events highlight the great extent to which the fusion of the changes in transport and communication affected the dynamics of interaction between various scientific and mathematical constituencies.

ELECTRICITY AND ELECTRO-TECHNOLOGY

The construction of the first electric dynamo by Michael Faraday in 1831 marked the start of the viable use of electricity in technology. Thirty years later, mathematical equations began to emerge out of the theory that described the link between electricity and magnetism, an insight that opened the door to applications in fields such as optics, wireless communication, power generation, and many more. Combined with the steam turbine and transmission technology, the ability to produce abundant electrical power and distribute it over vast distances transformed society.

George Green helped pave the way toward the mathematization of electricity and magnetism by showing how the notion of a potential function could be applied to describe electromagnetic phenomena. Green, who was 4th wrangler in 1838, published his seminal *Essay on the Application of Mathematical Analysis to the Theories of Electricity and Magnetism* in 1828, well before his formal Cambridge education; this was a remarkable achievement for a predominantly self-taught mathematician. His work was influenced by that of French mathematician, Simeon Poisson, and complimented similar, parallel developments by Carl Friedrich Gauss in Germany. Green's modesty prevented him from attempting to publish his *Essay* in any recognized journal, so only a handful of privately produced copies emerged at the time. Not until Thomson discovered Green's work in 1845 and then succeeded in republishing the *Essay* in the 1850s in one of the most important mathematical journals of the day, *Crelle's Journal*, did it gain the profile and circulation it deserved; it would not be republished in Britain until 1871. Thomson also shared Green's insight with the likes of Stokes, Helmholtz, and Maxwell, who all went on to discover applications for the theory. A useful summary of the evolution and application of the mathematics describing electromagnetism throughout the nineteenth century can be found in the third edition of volume 1 of Maxwell's *Treatise*, which was published in 1891 with an updated preface by J. J. Thomson.

Wired telegraphy developed through the 1830s, the first "electric telegraph" connecting Göttingen's Institute of Physics with the Astronomical Observatory in 1833. In Britain, a similar system was introduced on the railways to provide and automate signaling services. Routing transmission cables across land was not always easy but attempting to connect two continents separated by a vast sea was a challenge of a different order. Thus it comes as no surprise that the Atlantic Telegraph Cable project—the laying of a cable between the west coast of Ireland and Newfoundland—is often described as one of the greatest technological achievements of the nineteenth century (for details see Chapter 5). It took over twelve years, the equivalent of millions of dollars of investment, and in excess of 50 transatlantic voyages to achieve final success. The breakthrough in the end was largely due to the mathematical acumen of Thomson, who had been alerted to the project in 1854 by Stokes. The latter, along with others, had realized that laying an electric cable under water required a much greater understanding of electrical theory than that needed to lay a cable under or over the ground. For example, in 1857, when the cable broke unexpectedly, Thomson provided the mathematical explanation, using differential equations to analyze the mechanical forces involved in dropping such a cable to the bottom of the sea from the water's surface. With cable laid, information could be transmitted across the Atlantic within a matter of minutes as opposed to the ten days it would previously have taken by ship. Admittedly the volume was modest—the 1866 cable was initially capable of transmitting only eight words per minute—but the increase in efficiency was enormous.

Once Maxwell had provided a mathematical formulation that described the interplay between electrical and magnetic fields, a natural next step in communications was to attempt to dispense with the physical wires that connected the transmitter with the receiver. It was the British mathematician and physicist Oliver Heaviside who in 1884, using vector notation, restructured Maxwell's twenty equations into the four equations known today as "Maxwell's Equations." These four elegant statements summarize the mathematical theory underpinning every aspect of wireless communication that incorporates electromagnetism. Such is their fundamental importance that in 1964 the celebrated American theoretical physicist and Nobel laureate, Richard Feynman, wrote:

> From a long view of the history of mankind – seen from, say, ten thousand years from now – there can be little doubt that the most significant event of the 19th century will be judged as Maxwell's discovery of the laws of electrodynamics. The American Civil War will pale into provincial insignificance in comparison.[1]

In 1901, Guglielmo Marconi proved that wireless waves are not affected by the curvature of the earth by conducting the first successful transatlantic radio communication. This achievement followed decades of innovation in wireless telegraphy, a story punctuated with names such as Heinrich Hertz who was the first to prove the existence of electromagnetic waves predicted by Maxwell's

Equations, and Nikola Tesla, a flamboyant and innovative experimenter who made important contributions to the design of the modern alternating current (AC) electricity supply system that he demonstrated at the Chicago World's Fair. Two other influential protagonists in the electrical realm at the time were Oliver Lodge and Alexander Popov. Lodge performed much of his work as professor of physics and mathematics at University College Liverpool, whilst Popov used his time as a lecturer in electrical engineering and mathematics at the Kronstadt Torpedo School in his native Russia in similar fashion. Both men fed off the work of Hertz. Lodge paid homage to him at the BAAS meeting of 1894 in Oxford in a memorial lecture that included a practical demonstration of the quasi-optical nature of radio waves. Popov made use of the growing ease of travel to visit cities such as Berlin, Paris, Zurich, London, New York, and Chicago, where he interacted with the main players in the field, spreading news of his progress towards perfecting the transmission and reception of radio waves. He was rewarded in 1910 with an appointment at the St. Petersburg Electrotechnical Institute as professor of physics. There he set up a wireless telegraphy research laboratory, a facility that witnessed some of the first attempts at wireless voice communication.

Just as the changes in the efficacy and availability of transportation were making the world a smaller place regarding the movement of physical objects, so the globe was shrinking due to the ability to move information across it using electromagnetic waves. During the late 1870s, and early 1880s, progress was being made on both sides of the Atlantic towards realization of the first practical vacuum tube, the thermionic diode, a component that would have significant implications in the areas of communication and computing. The device, invented by John Fleming, who first learnt mathematics at UCL under Augustus De Morgan before going up to Cambridge, could convert alternating current into direct current at the scale required to integrate into electrical circuits. This was a major step forward and led immediately to the realization of more advanced transmitting and receiving equipment and, ultimately, to the wider field of electronics.

Advances in lighting technology came from the growing understanding of the nature of electricity. Hertha Ayrton, who studied mathematics at Cambridge from 1877 to 1881, made her first major contribution to engineering by inventing a drawing instrument that could divide a line into any number of equal parts. She subsequently spent much of her time studying the behavior and characteristics of electric arcs, and wrote an extensive treatise on her research, *The Electric Arc*, published in 1902. Shortly afterward, she had the distinction of becoming the first woman nominated for election to fellowship of the Royal Society, although her candidature was quashed on the grounds of her being a married woman, since married women had no status in law! Ayrton's work exemplified the emergent contribution of women mathematicians and scientists to the furtherance of technology, despite societal pressures to the contrary.

AERONAUTICS COMES OF AGE

The mathematics of Archimedes combined with eighteenth century materials and construction techniques proved sufficient resources to build the first balloons that lifted men and women into the air. Building heavier aircraft for controlled, fixed-wing, powered flight presented major new obstacles that would require a much deeper fusion of scientific disciplines and technological know-how. This was a natural evolution that played out through the whole of the nineteenth century, culminating in the now famous flights of the Wright brothers on December 17, 1903. Thereafter, a frantic race ensued to develop faster, stronger, safer, more agile and longer-range machines. Driven initially by industrialists seeking profit and notoriety, and promoted later by the more exacting demands of the military, the first two decades of the twentieth century saw escalating growth in almost every aspect of aeronautics. The concomitant burgeoning of the mathematics and engineering underpinning this growth is a tale littered with the names of often poorly acknowledged individuals whose efforts fueled and sustained these developments. The physical manifestation of the aircraft that appeared at the dawn of the 1900s was not the result of one idea or a particular piece of technology. The new machines emerged rather from the conflation, adaptation, and maturing of many disparate aspects of engineering, mathematics, and material science, all forged during the previous century (Royle 2020).

These efforts date back to the early 1800s with the work of Claude-Louis Navier and George Gabriel Stokes, who broke new ground in their studies of the general movement of fluids. This led to the famous Navier-Stokes equations, which have applications to many problems in physics and engineering, including models for predicting weather patterns, ocean currents, or the air flow around a wing. The investigations of Navier and Stokes paved the way for Edward Routh, who unwittingly laid the foundations of a mathematical description of the nature of the motion of a body in the air. Routh was a Cambridge Senior Wrangler, who famously beat Maxwell into second position in 1854, and he went on to become an extremely successful coach for numerous winners of the Tripos competition. One of his students, George Bryan, built upon his work in a study entitled *Stability in Aviation* of 1911, which introduced mathematical equations that give necessary conditions for stable flight. This was a significant advance since a stable aircraft will have a tendency in flight, when disturbed by some atmospheric phenomenon or pilot control input, to naturally return to its original state, rather than enter some sort of divergent oscillation or departure from controlled flight. In the vanguard of the quest to understand the nuances of the airfoil were several prominent mathematicians, scientists, and engineers, exemplified in Britain and Germany by the likes of Lord Rayleigh, Frederick Lanchester, and Ludwig Prandtl (Eckert 2019).

Rayleigh, Senior Wrangler in 1865, was a pivotal figure in contemporary science and mathematics in Britain for many decades. His personal interest in aeronautics began as early as 1891 when he wrote a review in *Nature* of *Experiments in Aerodynamics* by Samuel Pierpont Langley. In 1900, Rayleigh delivered the Wilde lecture on *The Mechanical Principles of Flight*. This enhanced his profile and, combined with his expertise in hydrodynamics, gave him the perfect résumé to justify his reputation as the driving force behind the discontinuity theory of airfoil lift, which was championed in Britain until the post-war era. The discontinuity theory's origins dated back to Rayleigh's work in 1876 when he studied the resistance of fluids. In experiments directed on a flat lamina at an angle to a water flow, he observed a region of dead water behind such a lamina that was bounded by a surface of discontinuity. Along this surface the pressure was the same as if there were no obstacle, which led Rayleigh to conclude that there must be an augmentation of pressure on the front face of the lamina.

The competing circulatory theory was associated with the German engineer Ludwig Prandtl. As it transpired, this theory offers a much closer representation of the underlying processes involved in generating airfoil lift. Prandtl, who became director of the prodigious Institute for Technical Physics at the University of Göttingen in 1904, later worked with fellow German engineers Albert Betz and Max Munk to develop a mathematical model capable of predicting the lift distribution over an airfoil in flight based on its geometry. The circulatory theory of lift argued that the presence of an airfoil in an airflow induced a circulation of air around it that modified the air's normal path. This reduced the pressure on the top surface of the airfoil whilst increasing it on the lower surface. The theory has since become known as Prandtl's "lifting line theory," although Lanchester had independently presented this idea in 1907, so that some authors call it the "Lanchester-Prandtl wing theory."

Frederick Lanchester had a less auspicious academic heritage than many of those who punctuated the early technical development of British aeronautics. He attended Finsbury Technical College in London, but failed to gain any formal qualifications despite his natural talent for engineering. This lack of official recognition did not prevent him from becoming an extremely successful businessman in the automotive trade, but it had serious ramifications for his credibility amongst academics, particularly those from Cambridge. Ironically, Lanchester had been arguing the merits of a circulatory component in theoretical lift analysis for some years. He postulated such a mechanism in his book *Aerodynamics* of 1907, but contemporary experts largely ignored it. Lanchester's perceived lack of mathematical rigor, coupled with his engineering background, worked against him, whereas Rayleigh's theory seemed more plausible to those who wielded most influence in British aerodynamics at the time.

Experimentation played a major role in early aerodynamical research. Wind tunnels thus sprang up in the major research centers across Europe and the United States in order to gather data from scale models of aircraft. Mathematics was then required so that the effects noted in these experimental facilities could be translated meaningfully to full-scale aircraft. Work was also going on to develop internal combustion engines of sufficient power to haul the rather basic early airframes into the sky. Achieving sufficient power-to-weight was one of the most pressing challenges facing the first cohort of aeronautical engineers. In Britain, efforts towards research and development were concentrated at two main facilities, the Royal Aircraft Factory, and the National Physical Laboratory (NPL).

The latter facility, which opened in 1900, set standards in science and engineering much like the Physikalisch-Technische Reichsanstalt (PTR) in Berlin. Various rumblings and statements of intent had been voiced at a number of BAAS meetings during the 1890s as unease grew over Germany's advantage in this field. The likes of Werner von Siemens and his successors had been driving Germany forward for well over a decade, and it was time for a British response. Thus the arrival of the NPL, the British equivalent of the PTR, established in the mid-1880s, was considered by many as long overdue. Initially the technical institutions and societies dictated the specializations at the NPL, where the chosen demarcations were: mechanics and engineering, electricity, optics, chemistry, metrology, terrestrial magnetism, and thermometry. These departments corresponded quite closely to a number of professional trades associations founded over the course of the nineteenth century, namely, the Institutions of Civil Engineers (1818), Mechanical Engineers (1847), and Electrical Engineers (1889). These organizations were, no doubt, most influential in determining the NPL's initial technical structure and focus. As time passed, however, its remit broadened and, in 1909, the Aeronautical Division appeared led by British mathematician Leonard Bairstow.

The Royal Aircraft Factory was born out of a Government-sponsored military ballooning project that had been set up in 1878. During the First World War, this establishment eventually attracted some of the finest scientific and mathematical minds in Britain together to tackle daunting aeronautical challenges. The main group of these pioneers became known as the Chudleigh Lot (after the name of the house they shared). In their pursuit of knowledge, many of them flew as pilots or observers in experimental aircraft. Typical of this group was Cambridge-educated Geoffrey Ingram Taylor, who spent much of his time studying airflow over airfoils in flight; he also used mathematics to inform the design of parachutes. Another pioneer was Frederick Lindemann who, along with George Paget Thomson and Hermann Glauert, derived the mathematics that described an aircraft spin. Not only did Lindemann tackle the mathematics, he actually took to the air himself to prove his analysis of the

FIGURE 6.7. Royal Aircraft Factory F.E.2b fighter of No. 23 Squadron Royal Flying Corps, 1916. Photo by Hulton Archive/Getty Images.

actions necessary to safely recover from such a disorientating and hair-raising maneuver.

In addition to the power-to-weight issue, the challenges of aviation were understandably many and varied at this time. Building airframes that could withstand the aerodynamic and other forces they would encounter was a perpetual concern. There were so many unknowns and aspects of flight to think about—lift, drag, aero-elasticity, stability, control, and so forth—that the problems seemed almost intractable. It can truly be said that to design and construct a new model of aircraft that had a reasonable hope of holding together during all phases of flight was a precarious and imprecise science. There was also a dichotomy of philosophy as to the best way to tackle these nagging issues. The more practical engineers favored progress at the expense of prevarication: build it, fly it, and see what happens. The more academically minded, on the other hand, preferred attempting to get at least some of the mathematics right before launching human beings into the air. Bryan, who was one of the most persistent advocates for mathematics, made the point:

> But when *Stability in Aviation* was in the press [1911] fatalities occurred daily, and killing off the pilots of unsafe machines was the only method that the practical man would have anything to do with. To invoke the assistance of a mathematician would have been an idea too terrible for words, and as for compensating him for his loss of time over the work, this might have cost £100, which would have been a preposterous waste of money when the same thing could be done by smashing up ten machines costing £1000 each.[2]

Aeronautics was a discipline that influenced technology both intrinsically and as a catalyst. The aircraft themselves were a manifestation of technological change, and yet they also served as vehicles to stimulate progress in science and technology. The eclectic nature of those comprising the Chudleigh Lot fostered an environment where multidisciplinary influence helped individuals to progress ideas in their own particular areas of expertise; Francis Aston was a typical example. He had been working at the Cavendish Laboratory in Cambridge prior to his appearance at the Royal Aircraft Factory and after the war he invented the mass spectroscope, using it to isolate stable isotopes of various elements. It was work that led to the award of a Nobel Prize in 1922. George Thomson would also receive a Nobel Prize, in 1937, for his work on the wave-like properties of the electron (the subatomic particle discovered by his father, J. J. Thomson, in 1897, and who had been awarded a Nobel Prize in 1906). Later in his career, Geoffrey Taylor would emphasize his technical credentials as an important member of the British contingent that formed part of the Manhattan Project to develop the first atomic bomb.

In many ways, the aircraft that engaged in combat during the First World War not only defined the state of a new engineering discipline, they summarized the state of technology and the influence mathematics had in attaining it. A typical machine was the Geoffrey de Havilland-designed F.E.2b, which was generally fitted with a 160 horsepower engine and capable of carrying external bombs and a cockpit-fired Lewis machine gun. It could also operate in the reconnaissance role fitted with an air-photography camera, and as a night fighter when armed with twin Lewis guns separated by a searchlight that drew power from an engine-driven generator.

Aircraft were vehicles that moved in three dimensions of space under well-defined laws of motion. They were powered by state-of-the-art internal combustion engines, and were constructed from the lightest and strongest materials available. Some aircraft were fitted with wireless communication devices in the form of spark-gap transmitters, which relied upon electricity and the propagation of electromagnetic waves. Some used powerful lights for communication. Reconnaissance aircraft carried robust and effective cameras to image the ground below. The quest to understand the behavior of airfoil wings pushed the envelope of fluid dynamics to its limits, and mathematics explained maneuvers such as the spin and the stall, and could be used to calculate the "g-forces" experienced by the pilots and airframe. A darker application of mathematics was in calculating the parameters for the trajectory of munitions, and in guiding the design of the bombsights used by aircrew in flight to define dropping points. The corporate experience gained by many of the mathematicians and scientists engaged in research and development on both sides of the conflict would be used to drive the various manifestations of technology to the next level in the post-war era. Some remained in aeronautics whilst others returned to their true vocations.

The end of the First World War signaled the end of a period during which the landscape of technology had changed beyond recognition. The world had become a much smaller place due to the innovations in transport and communications, and humankind's ability to contextualize its status in the universe had taken a leap forward. Behind all of the technological advancements and cognition was mathematics in some form or other, be it the theory that informed design and performance, or in the practical instruments that captured ever more valuable and accurate data. Progress had been driven by a fusion of pure and applied mathematics that had manifested itself as practical engineering. Electromagnetism had been the new kid on the block, but had established itself as a valued resident on the technological estate. Aeronautics was poised to bring commercial and leisure travel to the masses. The seeds had been sown for the computer age to begin and, as the implications of Einstein's equations were still being absorbed, it was clear that mathematics had helped bring humanity to the start of a new and exciting era of discovery.

CHAPTER SEVEN

Representing Mathematics

Material Practices of Producing and Visualizing Knowledge

ANJA SATTELMACHER AND DAVID E. ROWE

The word "representation," as we learn from the *Encyclopaedia Britannica* in its eleventh edition from 1911, comes from the Latin word "re-presentare," to bring into presence or to make present again. The English cognate appeared for the first time in English dictionaries of the thirteenth century, whereas in the sixteenth the term typically referred to the performance of a play. By the late seventeenth century, it entered psychology to describe the action of mental reproduction, though in a technical sense that applied especially to the immediate object of imagination. A century later, the notion assumed a central role in Kantian philosophy as "representation" (*Darstellung*) became the generic term for percepts, concepts and ideas. Finally, it came to mean "substituted presence" as opposed to "actual presence," from which its political meaning derives—a representative person who as a single individual speaks for a larger group (e.g., in Parliament).

When using the term representation in connection with mathematics, one must also carefully specify the context in order to understand the meaning of the word. In this chapter, it will refer mainly to a rather early meaning, namely "to show" or "to demonstrate" something. Throughout the nineteenth century, representing mathematics usually meant connecting mathematical knowledge with pedagogical practices. This linkage was natural, first, owing to the imminent meaning of representing—showing—as this related to the teaching of mathematics at colleges and universities, as well as at primary schools. Second, it reflects how the spread of new approaches to teaching geometry on the continent of Europe took place against the background of new pedagogical ideas and concepts, which had emerged during the last decades of the eighteenth century.

Alongside these trends, there arose a new interest in grasping and visualizing geometrical relationships immediately rather than by way of analytical methods. During the early nineteenth century, French geometers, led by Gaspard Monge, revived synthetic geometry in a more dynamic form that eventually developed into modern projective geometry. Its starting point, however, was the study of concrete objects in space and their properties under projection. One of Monge's students, Théodore Olivier, used similar methods to go a step further. Whereas Monge's methods reduced spatial relations to two-dimensional drawings, Olivier applied them to produce three-dimensional models. His movable string models became justly famous and found their way into collections even in the United States. This new French approach to synthetic geometry had a broad impact on the Continent, though mainly outside institutions that promoted humanistic education. It left few traces in Great Britain, where the longstanding veneration for classical synthetic geometry continued to dominate in the British schools. This contrasts strikingly with the situation on the Continent, where already during the eighteenth century Euclid's *Elements* had nearly disappeared (see Volume 4).

Many of the standard objects investigated by French mathematicians were already familiar, though primarily in connection with their algebraic or analytic properties. For example, mathematicians since the time of Leonhard Euler knew how to classify the different types of quadric surfaces by means of their algebraic forms. By 1800, these second-degree surfaces became a standard part of analytic geometry, which developed techniques for classifying geometrical objects. Thus, second-degree equations in two variables corresponded to the various types of conic sections in a plane (ellipse, hyperbola, and parabola). Using Cartesian coordinates, the points on quadric surfaces satisfy a second-degree equation in three variables; they are, therefore, the analogues of the conic sections and their classification leads to five different types. Several of these are ruled surfaces, objects generated kinematically as families of lines in space. Others, such as an ellipsoid, do not contain any lines; in such cases, one often represented the surface as the boundary of a solid body.

These investigations led to a deeper understanding of spatial relationships, for example, by representing curves that lie on quadric surfaces, which contain families of conic sections. The focus in these studies fell on geometrical properties with relatively little attention to algebraic relationships. A geometrical model then served as a material representation for an abstract object ideally situated in a three-dimensional mathematical space. This does not mean that Monge neglected analytical methods, far from it. His approach to synthetic geometry, which he incubated with the founding of the École Polytechnique (EP) in 1794, supplemented the heavy course load in analysis and mechanics (see Chapter 2). Monge's courses in descriptive geometry were especially influential. In them, he taught future engineers the basic methods of scientific drawing, which used orthogonal projections to derive data needed for designing a building. These practical techniques eventually spread throughout Europe and even beyond (Barbin, Menghini, and Volkert 2019).

Institutional factors played a major role when it came to this reception of French geometry in the German states. From around mid-century onwards, a number of polytechnic institutes emerged in German-speaking countries. These colleges had a lower status than the traditional universities, though they grew in importance over the course of the century when technical education became part of the general effort to promote modern standards of industrialization. The Mongian teaching tradition established at the École polytechnique had a major impact on several of the leading German polytechnic schools, several of which cultivated drawing and model making. The polymath geometer Christian Wiener founded one of the first large collections of mathematical models at the polytechnic school in Karlsruhe, for which he purchased some of the models from the Olivier collection in Paris. In the mid-1870s, Wiener's nephew Alexander Brill founded a modeling workshop at the Munich Technical College, and soon afterward, he launched a business selling models through his brother's publishing company. Meanwhile, in their courses at German universities, geometers generally neglected the kinds of practical problems that arise when teaching students methods for drawing curves or building three-dimensional models.

A major exception in this respect, however, was Felix Klein, who promoted building and collecting models wherever he taught (Tobies 2021). As the most influential mathematics instructor in Germany during the 1880s and early 1890s, Klein also exerted a major impact on mathematics in the United States (see Chapter 2). In 1893, he took part in the Mathematical Congress held at the World's Columbian Exhibition in Chicago, where he gave demonstrations of the mathematical models on display there (Kidwell 1996; Kidwell, Ackerberg-Hastings, and Roberts 2008; Parshall and Rowe 1994: 305–9, 338–41). This unusual performance suggests how the term "representation" in this context has a twofold meaning. In the purely mathematical sense, geometrical models represent a material embodiment of a mathematical formula. Yet, in a second, less literal sense, they were professional status symbols for their scientific persona. Mathematicians like Klein could point to models and instruments as a visible expression of knowledge that would otherwise have remained invisible. These artifacts thus took on strategic importance for Klein as part of his efforts to expand the role of mathematics in German scientific culture by bridging the gulf that separated the universities from the technical institutes.

PRACTICE, PEDAGOGY, AND REPRESENTATION

An important concept relating to the word "representation," and which accompanied mathematical thinking throughout the entire nineteenth century, was mathematical intuition ("*Anschauung*"). This referred to a mode of visualizing that went beyond the immediate sense of sight and included the haptic sense as well. This term intuition/*Anschauung* had deep roots in philosophy and in pedagogy dating back to the eighteenth century. When used by German mathematicians in the next century, this usually meant a type of intellectual activity they could teach with the aid of appropriate tools, such as models or drawings.

Among the many ideas that emphasized visualization as well as an intuitive approach to learning was the concept of sensualism, which spread from England and France to other European countries in the second half of the eighteenth century. Among its leading advocates were John Locke, Etienne Bonnot de Condillac, and Jean-Jacques Rousseau. Starting from the premise that the senses are fundamental to the attainment of knowledge, all three placed human experience at the center of their considerations. Especially influential for pedagogy in Germany was Rousseau's widely received *Emile*, written in 1762. Many leading educators identified with his idea of the free development of a child's personality through the formation of an inner sense. According to Rousseau, the optimal way to promote a capacity for abstract thinking in children was by building on their experience with tangible objects, so-called "*realia*" (Whitmer 2015). These sensualist ideas spread within enlightened pedagogical circles throughout the late eighteenth and early nineteenth centuries.

Among the recipients in Germany were the so-called philanthropists, who introduced elementary education that aimed to promote learning through observation, practical work and self-activity. Their leaders founded model schools with curricula focused on reading, writing, and arithmetic, but also offering a wide variety of other subjects as well, including natural history, geography, geometry, astronomy and physics, personal hygiene, and moral education. The best known of these model institutions was the *Philanthropinum*, founded in Dessau in 1774 by the Hamburg educator Johann Bernhard Basedow. This school played a pioneering role for modern school reforms by adopting educational principles that embodied the spirit of the Enlightenment. Basedow based his philosophy of learning on "intuitive thinking" (*anschauendes Denken*), which aimed to direct children's sensory experiences in such a way that they gained an understanding of the world at an early age (te Heesen 2002). A standard method from this time was to make use of clever arrangements of images in child-oriented publications. By so doing, according to this theory, one could enhance a child's thinking capacity and make it more malleable.

Two influential educators who followed in the same direction were the Swiss Johann Heinrich Pestalozzi and his German counterpart Friedrich Wilhelm August Fröbel. Both took sensory perception and the promotion of visual cognition as the starting point for their respective educational practices. Drawing on Rousseau and sensualism, Pestalozzi wrote a large number of elementary books based on his pedagogical experiments, in which he stressed spelling, speaking, reading, writing, calculating and drawing. Like the philanthropists, for Pestalozzi the principle of development was decisive: elementary education should start with the simplest matters and build from there. In these texts, he called visual instruction the "key to experience and knowledge." The untrained child should not be confronted with points and lines at the very beginning of school—this, according to Pestalozzi's ABC of mathematical perception for mothers. Instead, a child's perception should first focus on concrete objects from everyday life and then slowly confront the realm of abstract ideas.

Following the lead of Rousseau, the philanthropists, and Pestalozzi, Fröbel promoted the idea of the "active child" and "self-sufficient education." In the 1830s, he developed play materials designed to help children grasp basic objects of spatial intuition, especially elementary shapes such as spheres and cubes, but also the distinction between points, lines, surfaces, and solids. His aim was thus to begin with vivid objects and gradually progress from the simple to the difficult, thereby laying the foundation for abstract conceptual thinking and awareness. Other teachers throughout Germany took over some of the sensualist and philanthropic ideas by introducing a two-step learning process in mathematics lessons that included the systematic use of teaching aids as well as the study of abstract mathematical facts (Friedman 2018: 206–16; Sattelmacher 2021: 62–72). Formal structures—such as equations, axioms, formulas and rules—came only after patient preparation. This general pedagogical approach took many forms, including the use of teaching aids, such as models or drawings. These aimed to strengthen an intuitive feel for mathematics; they acted as an instrument for gaining an initial understanding of more abstract ways of thinking.

Mathematics and mathematical thinking have always evolved within the larger intellectual modes of the time. This was especially true for the European educational systems and social structures of the nineteenth century. As discussed in Chapter 1, most school instruction in mathematics aimed to impart arithmetical skills and left geometry out of the curriculum until the final years. Only an elite group learned the rudiments of solid geometry, as the focus was normally restricted to elementary geometry in the plane, which emphasized constructions using only straightedge and compass. Various other forms of *mathematical practice* were wide ranging, however, just as this term has many different meanings (Ferreirós 2016: 28ff). The present chapter focuses on practices connected with the production of geometrical models, a rather small corner of activity that nevertheless became significant and highly visible over the course of the nineteenth century. As didactical tools, these models aimed to promote the visualization of geometric relationships, though it appears doubtful that these didactic objects ever found systematic use in mathematics courses.

MODELING IN THE FRENCH TRADITION

The turn of the nineteenth century marks an important milestone in the history of practical mathematical education that would shape the scientific landscape for the next hundred years and beyond. As noted in the introduction to this volume, the French Revolution resulted in a rapid transformation of continental European science as a whole. With the founding of the EP, France initiated a trend toward professionalization in engineering mathematics that served as a model for a number of polytechnic schools later founded in German-speaking states. These developments went hand-in-hand with the (somewhat delayed) impact of the Industrial Revolution. This fed international competitiveness and a resulting demand for resources promoting technical and industrial education. Industrial progress required engineers with the necessary arithmetical, drawing and modeling skills, training they could gain from courses taught at the newly founded polytechnic schools. Following the program designed by Gaspard Monge at the EP, many of these schools in Germany introduced courses in scientific drawing that aimed to provide a complete systematization and quantification of practical knowledge. This tradition was later cultivated at the Zurich Polytechnic by the geometer Wilhelm Fiedler (Volkert 2022).

More than three decades after the founding of the EP, a new impulse in this direction came from the École centrale des arts et manufactures, which forged new connections between science, technology and industry (Alder 1999). Among its founders in 1829 was Théodore Olivier, one of Monge's first students. This convergence of modern interests eventually led to a new discipline, "industrial science," built on a strong alliance between theory and practice (Fox 2012: 9–18, 36–7). France was facing economic and technical challenges in the wake of the industrial revolution, creating the need for civil engineers. The École centrale met that demand by providing practical as well as scientific training for engineers and architects who would work with a variety of new building materials, ranging from iron to glass and concrete. They took courses on mechanical engineering, industrial and railway construction, but also in aspects such as the technical safety of structures and public hygiene with a direct bearing on social issues relating to improved living standards for urban life. The scope of scientific training covered fields that included metallurgy, industrial physics, analytical chemistry, mineralogy, geology, and descriptive geometry. During their first year, students took courses in mathematics, physics, mechanics, and chemistry, each with laboratory work, along with geological fieldwork. At the end of the year, they took a design examination in which they had to explain how to construct a factory machine.

Olivier's courses in descriptive geometry at the École centrale adopted a didactical principle dating back to Monge, who emphasized the importance of visual representations, in particular scientific drawings, to elucidate applications of mathematical analysis (Brechenmacher 2022). Olivier went a step further by employing three-dimensional models in order to represent mathematical concepts within geometry. Like his former teacher, though, he used orthogonal projections to represent spatial objects, for which he devised a special method for teaching the "description" and the "reading" of points and lines. He called this method *alphabet of the point* ("alphabet du point"), which he used to specify the spatial position of points or lines.

From transcripts of his courses from the 1840s, it appears Olivier conducted exercises with students at the École centrale, in which he taught them how to use this point alphabet. It was probably in this context that he used models to demonstrate these principles, though we cannot be sure, since neither he nor his colleagues or students ever mentioned their actual use during Olivier's lifetime. His string models, on the other hand, aimed to convey the properties of various types of ruled surfaces, which played a prominent role in his courses in descriptive geometry (Sattelmacher 2021: 78). A standard example from architecture involves finding the curve of intersection for two cylindrical surfaces.

Unlike most geometrical models, many of those designed by Olivier were not static objects. Using copper frames and silk strings, he constructed a new type of dynamic model that could represent whole families of ruled surfaces. Shortly after founding the École centrale, Olivier presented his first self-made model to the public—representing the lines on a special type of quadric surface representing a hyperbolic paraboloid. This model consisted of a frame with two V-shaped copper strips connected at their lower ends by two hinges mounted on a two-legged stand (see Figure 7.1). To this resulting skewed frame, Olivier attached a network of silk threads by knotting them to the perforated metal frame. These strands were in two groups with two different colors forming a crisscross pattern; thus, only threads of unlike color crossed one another. Olivier used those of the same color to represent one of the two families of lines that lie on this ruled surface. To hold the threads taut, he attached lead weights on the ends of each. By slowly moving the opposite side of the frame, this model could represent an entire family of second-degree surfaces of varying curvature.

Olivier's purpose was thus to show how to deform ruled surfaces by altering the position of the threads of the model. In the case of the hyperbolic paraboloid, one accomplished this by gradually changing the angle of inclination of the two pieces of its frame at the hinge connecting them. For several of his more elaborate models, Olivier built special mechanisms for deforming the strings. Some showed pairs of hyperboloids, with one set of generating lines each represented by threads in two different colors, each attached to two movable

FIGURE 7.1. Thread model of a deformable family of hyperbolic paraboloids in the style of Olivier. Photo courtesy of Gerd Fischer.

discs. By rotating these discs, one could either tighten or loosen the strings (see Figure 7.2). In their limiting positions, the threads of the inner hyperboloid would tighten until forming a cone, whereas the outer surface would loosen until the strings formed a cylinder (which one can view as a cone with its vertex displaced to infinity). In a critical intermediate position, the two surfaces merged, revealing the two systems of generators that lie on a hyperboloid of one sheet. A particularly impressive model shows how a cone with an associated tangent plane transformed into a hyperboloid with a tangent surface in the form of a hyperbolic paraboloid. In some cases, one could deform these figures by sliding, turning, or tilting the discs without thereby changing the qualitative features of the model.

During the 1830s, Olivier worked closely with the Parisian instrument maker Hippolyte Pixii, who manufactured and marketed many of the models he had

FIGURE 7.2. An Olivier model showing a hyperboloid with a hyperbolic paraboloid as a tangent surface, manufactured by Fabre de Lagrange. Photo by SSPL / Getty Images.

designed. Another Paris firm, Fabre de Lagrange, later acquired this business and continued to produce and sell Olivier's models long after his death in 1853. A large part of his personal collection found its way to Union College in Schenectady, New York, when a professor of civil engineering purchased a set from Olivier's widow. This collection and others outside France still survive today, testifying to widespread interest in these objects throughout the nineteenth century. A particularly important acquisition took place in 1872 when the newly founded South Kensington Museum in London purchased a large collection from Fabre de Lagrange to strengthen this aspect for its holdings in the industrial arts.

That collection is particularly significant because the British mathematician Charles Merrifield gave a detailed account of these models of ruled surfaces in a catalogue published by the museum. Merrifield was Principal of the Royal School of Naval Architecture and Marine Engineering at South Kensington, thus someone with keen awareness of the larger purpose of these models. In his description of their general properties, he emphasized how fragile the models actually were:

> The models are constructed with especial reference to the possibility of changing their shape, by moving some of the supports of the strings, by altering the lengths or positions of certain parts, or by converting upright forms into oblique. [. . .] This possibility of *deformation*, as the process is technically called, greatly enhances the value of the models, by allowing them to represent a much greater variety of surfaces than if they were fixed. [. . .] They are, however, too delicate to be much pulled about, and, unless they are very cautiously handled, the strings are apt be become entangled or break.[1]

The models Olivier designed throughout his career—about 50 different ones altogether—had become widely familiar by the 1870s, when a strong surge of interest in model making began. They thus served as prototypes for much of this later activity, as reflected in the models found in collections in Germany, England, Italy, and many other countries. Many times, however, when model makers copied or redesigned standard types, they altered their materiality. Thus, while some companies relied on cardboard or plaster, others preferred wood or wire.

Although one often finds references to their role in didactics, namely as aids for enhancing geometric intuition, there seems to be little evidence that models played an important role in teaching. In most cases, those who engaged in producing models—using drawings to design them, tools to build them, and formulas to describe them—were the prime beneficiaries. Once built, however, they seldom became objects for further study. This suggests that their main didactical effect was on the model builders and not those who merely contemplated a finished product.

MODEL MAKING IN GERMANY

It is generally very difficult to find primary sources that provide detailed information about the practices of model making during the nineteenth century. An exceptional case, however, involves the mathematician Alexander Brill, a well-known maker and collector of mathematical models. Late in his life, he composed an autobiography that offers many insights into the culture of model making (Seidl, Loose, and Bierende 2018; Sattelmacher 2019). Brill was a native of Darmstadt and held a professorship at its Technical College until 1875. In that year, he joined Felix Klein at the Technical College (Technische Hochschule or TH) in Munich, where together they founded a "modeling cabinet" in which students learned how to design and construct geometrical models (Rowe 2013). Klein's interest in models and *anschauliche Geometrie* dated back to his student days under Julius Plücker in Bonn, as his mentor had already spearheaded this activity in the 1860s (see the Introduction).

Nevertheless, it was Brill and not Klein who played the leading role in supervising this hands-on work in Munich.

Alexander Brill was the oldest of six children. His father, Heinrich Konrad Brill, owned a bookshop and printing business in Darmstadt, later taken over by his brother Ludwig. His mother, Julie Henriette Simonetta Wiener, was an older sister of Christian Wiener, who for nearly a half century, starting in 1852, taught descriptive geometry at the Karlsruhe Polytechnic. According to Brill's own testimony, his uncle advised him to study architecture before turning to mathematics. Brill's wife, Anna Schleiermacher, also came from an established Darmstadt family with diverse scientific interests; it was mainly through her that Brill became interested in arts and crafts. Like many other model makers in Germany, Brill was involved in a dense network of personal relationships with like-minded researchers. His uncle already gained inspiration for constructing geometrical models in 1851 when he attended the Crystal Palace Exhibition in London. Later, his Karlsruhe colleague, Alfred Clebsch, encouraged him to construct a plaster model showing all 27 lines that lie on a cubic surface. This particular artifact stirred great interest when Wiener first exhibited it in 1868 at a gathering of about twenty mathematicians in the Bergstrasse, south of Darmstadt; afterward, he made and sold copies to satisfy the demand for it. His son, Hermann Wiener, later became a prominent model maker in Darmstadt, where he taught geometry for over thirty years.

Many of Brill's family members, including those on his wife's side, worked outside academia in various commercial and technical enterprises. These extra-academic connections proved to be an indispensable resource for his work as a model maker, an interest that began as a hobby. Brill's career fits the mold of a nineteenth-century craftsman-scholar (Sibum 2003) or that of the hybrid scholar (Klein 2012). These were personalities whose work fell into various niches located between the sciences and older craft traditions. Although he was a full professor at a technical university, Brill acquired his knowledge regarding the production of models from others—plaster makers, woodworkers, or bookbinders—who worked outside academia. He then took this locally acquired knowledge back to the TH, where he passed it on to his students. Alexander Brill was familiar with some of the German companies that produced mathematical models and instruments, but he did not find their geometrical products satisfactory. In particular, he criticized the quality of their work when it came to representing the curvature properties of surfaces. Consequently, he used his own network of contacts to manufacture the models that his brother Ludwig would advertise and sell through the family company in Darmstadt.

A glimpse of Brill's early activities at the TH in Munich appears in a letter he composed in 1875. Shortly after establishing a modeling cabinet there, he wrote to his former colleague in Darmstadt, the geometer Rudolf Sturm:

> I had casts made of the models of the center of curvature surface of the elliptical paraboloid and of the surface of constant negative curvature with geodesic lines [. . .] for duplication. You can obtain these later from my brother in Darmstadt. The models are delicate. I also had my brother-in-law August copy Plateau's experiments. The minimal surfaces are produced by soap bubble film within wire frames, which my brother-in-law has formed in quite a number. You see, we are cultivating a kind of visual instruction ("Anschauungsunterricht") here, which is probably not practiced elsewhere.[2]

Joseph Plateau was a Belgian physicist, who only two years earlier had published results from his experiments in which he used soap bubbles to study surfaces of minimal area that span a given frame. This work laid the groundwork for what mathematicians would later call the Plateau problem in the theory of minimal surfaces. Thus, Brill's brother-in-law, August Schleiermacher, was working on the cutting edge of what would later become an important area of mathematical research.

These types of family connections played a major role in this new enterprise. This document reveals, in fact, that already in 1875, Brill's brother Ludwig was planning to market models that Alexander Brill and his students were beginning to produce in Munich. In the coming years, models from the Brill collection proliferated to educational institutions all over Germany, but also to leading mathematical centers in Europe and even beyond. In his autobiography as well as in some of his published articles, Brill described how he and his students at the TH conducted their work on models, which began with careful calculations and drawings made by the students.

Unlike Brill, Klein had relatively little experience when it came to building models. He had only assisted Plücker in designing them, but his mentor then outsourced the actual work to a company in Cologne. After Plücker's death in 1868, Klein continued to work with the same firm, but also with a friend who constructed models. During the 1870s, Klein focused most of his interest on modeling higher-order surfaces (especially cubics and quartics), whereas Brill devoted a great deal of his efforts toward producing models of mainstream objects, including quadric surfaces. Although these were already very familiar after 1800, he found new ways to model them using mainly cardboard and plaster as materials. In contrast to Olivier's elegant, but fragile models of quadric surfaces made with copper and silk threads, Brill's models were far more durable and much less expensive.

The first series of models sold by the Brill publishing house were made of cardboard, namely a set of six quadric surfaces: an ellipsoid, hyperboloids of one and two sheets, an elliptical and a hyperbolic paraboloid, and a cone (since its vertex is a singular point, this is actually a degenerate quadric). Brill's inspiration for building these came after seeing some similar models constructed

FIGURE 7.3. Cardboard models made by Rudolf Diesel during his time as a student at the Technische Hochschule München. Diesel later donated the models to the CNAM. Collections of the CNAM. Photo © Anja Sattelmacher.

by Olaus Henrici, a geometer who had studied under Christian Wiener in Karlsruhe. Brill likely contacted Henrici directly, as he had been charged with helping organize a temporary exhibition of mathematical models, which went on display during an informal meeting of mathematicians held in Göttingen in 1873. Brill and Klein worked together in planning this event as members of the organizing committee. Its larger purpose was to discuss plans for eventually launching a German Mathematical Society, though this idea turned out to be premature and had to await a renewed effort in 1890. The sudden unexpected death of Göttingen's Alfred Clebsch in November 1872 was a major factor that dampened enthusiasm for this plan. Even so, more than fifty mathematicians attended, including nine from foreign countries, and on the final day they heard Brill deliver a stirring speech in which he extolled the benefits of the kind of convivial social interaction they had enjoyed during the previous days.[3]

The participants also had the opportunity to meet in the quarters of the physics institute as the guest of its director, Wilhelm Weber. There they found an impressive display of models, including 22 Plücker models made in zinc, supplemented by four others designed by Klein, all produced by the Eigel firm in Cologne (Rowe 2022). Hermann Amandus Schwarz, then teaching at the

Zurich Polytechnic, sent models of minimal surfaces, including two plaster models of periodic minimal surfaces, a type he first uncovered in the mid-1860s. Adolf Weiler, who had worked closely with Clebsch and Klein in Göttingen, exhibited the models of cubic surfaces he had designed for both of them the year before. One of these later became famous as the Clebsch diagonal surface, a highly symmetric cubic on which the 27 lines form a distinctive pattern. The organizers had to report one major disappointment, though, namely a large collection of plaster models sent by the firm of Charles Delagrave in Paris did not arrive in time, probably due to customs regulations. However, they did receive an album of photographs of these models, which at least allowed the participants to gain an impression of the missing collection.

Since this meeting took place just two years after the Franco-Prussian War of 1870–1 ended, these signs of friendly scientific relations between parties from the two belligerent nations would seem to be not without significance. Even more startling, though, is the fact that the organizers were able to procure a collection of Olivier's string models for their meeting, though not from an institution in France. During the war, Prussian forces took control of the Military School in Saint-Cyr, located far to the west of Paris. There they discovered an impressive collection of models, which some officers decided to confiscate as war booty and send off to the War Academy in Berlin. Klein may well have learned about this collection, and then placed a request to have some of these models shipped to Göttingen as a temporary loan. In any event, he was keenly interested in the Olivier models. Shortly before the war had broken broke out, Klein was in Paris and went to visit the Conservatoire National des Arts et Métiers (CNAM). He wanted to examine the Olivier models firsthand, but came away disappointed, as those he saw there were in abysmal condition.

Olaus Henrici had recently assumed the professorship briefly held by Thomas Archer Hirst at University College London. Both he and Hirst were avid proponents of the physicalist tradition in geometry, which placed them at odds with the dominant Oxbridge tradition that continued to advocate teaching geometry along the lines of Euclid's *Elements*. Hirst had studied in Germany, taking his doctorate in Marburg in 1852. Henrici, as a former pupil of Clebsch and Christian Wiener, essentially imported the German tradition of model making into England. At University College, he established a workshop very similar to the one Brill and Klein founded in Munich. Its novelty was such that reports on it appeared twice in *Nature*, emphasizing how Henrici used a great variety of visual aids to support his teaching:

> to show the simple contrivances employed, we may say these models were made of knitting-needles with small spherical ends of sealing-wax of different colors, thus enabling the student to see their different directions; in other cases joins were indicated by ties of different colored wool, thus showing that the corresponding points of two perspective triangles meet in a line.

> Projective rows of points made of pricked wood, the corresponding points joined by India-rubber threads; models exhibiting the generation of ruled surfaces of the second order, movable models made of silk threads stretched by weights, parallel pencils of lines making the paraboloid. The generation of curves by the intersection of pencils of lines; this was shown by two flat pencils of lines made of colored silk in mahogany frames, one of which moved upon the other; at the intersection of certain pairs of threads were placed small indices which clearly showed to the eye various forms of ellipses and hyperbolas. (Barrow-Green 2021: 53)

The models Henrici sent over were nothing like smooth surfaces. Instead, he built them by wedging together semi-circles of varying sizes and in two parallel arrays, thereby using these cross-sections to evoke the shape of a quadric surface. Their truly novel feature, though, was that one could alter their shape since the model was deformable. Brill replaced Henrici's semi-circles with full circles and introduced other refinements, but he also continued to experiment with cardboard models in his Munich seminars, where his students made precise calculations or improvements with the material. An example of this led to the cardboard models designed by Brill's student Rudolf Diesel, who later became famous for his patented invention of the Diesel motor engine. His cardboard quadrics consisted of prefabricated cross-sections, which he slid together vertically (see Figure 7.3).

The mathematics underlying these types of models was quite elementary, but also rather laborious. It relied on calculations of the cross-sections using standard methods of analytic geometry. One starts with a simple second-degree equation in three variables for the given quadric surface together with an equation for an appropriately chosen one-parameter family of parallel planes. Each plane then intersects the quadric in a conic section. One finds equations for these conics by eliminating one of the variables in the original equations, which yields a one-parameter system of second-degree equations in two variables corresponding to the conics in each plane.

The experimental phase then began, since the construction required wedging together a large number of cross-sections from two families of parallel conics. Decisions about the ideal number or the simplest families of planes one might choose depended on practical experience. In the case of an oblique ellipsoid, only two families of parallel planes yielded cross-sections in which the ellipses were circles, which greatly simplified the overall construction. Once the calculations were completed, the next step was to draw these curves on pieces of cardboard and cut along these contours. In order to create a three-dimensional model from these two-dimensional templates, one made vertical slits in the latter wherever two of the cross-sections intersected. These marked the places where two corresponding pieces of the model were to be slid together. Once constructed, these cardboard models were very flexible and easy to handle. The ellipsoid with circular cross-sections described above was even collapsible, so one could flatten it out and place it in a box, then take it back out, tilt the model to make it fan back out, and place it on a base in a collection.

Cardboard models, such as these, proved relatively easy to produce at low cost since the materials involved did not require specialized technical or craft skills. They were and still are highly suitable for classroom use, since the only building materials required are pencil, compass, scissors, and cardboard. Still, the process of cutting and wedging the pieces together requires many hours of work, which makes such a project quite unsuitable for large groups. In Brill's day, it was far more practical simply to purchase the full set of six models, which were quite inexpensive.

Quadric surfaces, including the simpler types with rotational symmetries (cones, spheres, etc.) abound in model collections the world over, and yet there are subtle differences. These arise because the same type of mathematical object could be represented by models produced using various materials, which may then have different didactic, epistemic, or simply practical functions. Alexander Brill's cousin, Hermann Wiener, designed models exclusively built using wire (see Figure 7.4). This involved a division of labor that reflected a higher level of professional artisanship. Although Wiener made the drawings himself, he had a precision-engineer form the models according to his drafts. No students or family members were involved in this work, which Wiener ran as a conventional business rather than a cottage industry.

Olaus Henrici had experimented earlier with models of quadrics made from sticks. A well-known method for describing a system of generators that lie on a hyperboloid of one sheet begins with three mutually skew lines in space. Lines that intersect all three then form a system of generators (choosing any three of these determines the other system). In 1873, Henrici challenged one of his students at University College to construct a model based on this principle. He advised him, though, that the model would soon become rigid as he built it, but then learned to his own surprise that this was not the case. Once he realized his mistake, Henrici noticed that the deformations that leave the generators fixed lead to a system of confocal hyperboloids. Hermann Wiener later built and sold a wire model that illustrated Henrici's finding (Barrow-Green 2021: 70–1).

The choice of materials for producing models depended on many different factors. An ellipsoid made from plaster was highly stable, but also opaque, whereas the same object in cardboard was light and flexible, and when made from wire rods the hinges are visible. Olivier's thread models of ruled surfaces were aesthetic marvels, but also very fragile. Those built later by Fabre de Lagrange no longer had visible lead weights to hold the strings taut. Instead, the company usually mounted the model atop a wooden box-like stand with a hole into which the strings disappeared. Since the weights were now inside this opaque box, they no longer distracted from the visual effect conveyed by the model. As noted earlier, the threads on many of these models were movable,

FIGURE 7.4. Display of wire models from Herrmann Wiener's catalogue, 1912. Image courtesy of Anja Sattelmacher.

whereas plaster models only illustrated the properties of a single geometrical object, unlike cardboard models. The latter were in many ways ideal for classroom use, since they were lightweight and easy to package and transport, unlike plaster models, which were much more difficult to handle. These, on the other hand, had the advantage of being easier to reproduce and distribute to other collections. Wire, by contrast, had very different characteristics: wire models were well suited for projection onto a screen, a technique that became more common due to increasing numbers of students.

The models of greatest importance in the Brill collection, however, were those made in plaster, a material widely used for model practices throughout the century. The many elegant plaster models he and his students designed for this undertaking were decisive for its success. For Brill, the decision to use gypsum was mainly practical, since one could easily replicate models built with this substance. Moreover, he was in an environment where practical knowledge about handling this material was readily available. This was most fortunate, as it meant that Brill had access to artisans who could accentuate the aesthetic qualities of these mathematical artifacts, something he did not leave to chance. For although the prototypes were produced at the TH Munich, the casts for the models came from the workshop of Joseph Kreittmayr, a master craftsman who made plaster casts of antique statuary for the Bavarian National Museum. The students at the TH used a different modeling clay for their prototypes because plaster hardens very quickly and was therefore not easy to modify. Brill brought these master models to Kreittmayr, who then used them to produce negative molds for plaster casts. Thanks to this connection, the prototypes made at Brill's institute emerged in an elegant new form. From Darmstadt, the firm of L. Brill advertised and sold these plaster models with considerable success up until nearly the end of the century. An advertisement in the *American Journal of Mathematics* from 1890 indicated that the company's catalogue listed 217 models for instruction in higher mathematics, of which 158 were made in plaster, 40 in wire, and 19 using silk threads (Bierende and Loose 2018: 99). Indeed, the Brill models continued to dominate this niche in the production of mathematical models and instruments even after 1899, when the company sold its rights to the firm owned by Martin Schilling in Halle.

MARKETING, COLLECTING, AND EXHIBITING

Family ties were clearly an important factor in accounting for Alexander Brill's career as a model maker, though his case was by no means atypical. Martin Schilling's brother was the geometer Friedrich Schilling, who took his doctorate under Felix Klein in 1893. Klein was keen to promote his career, especially since his skills in descriptive geometry fit Klein's plans for expanding Göttingen's program in applied mathematics. After Arthur Schoenflies left in 1899 to assume a professorship in Königsberg, Klein arranged for Schilling to replace him as associate professor. Along with his teaching duties, Friedrich Schilling also took over responsibility for the Göttingen Collection of Mathematical Instruments and Models.

Thus, it seems hardly an accident that this appointment came in the very year that Martin Schilling acquired the rights to market the Brill models. Indeed, this symbiosis between the two brothers was to prove extremely fruitful. Friedrich had both a deep knowledge of models as well as contacts with mathematicians interested in buying them for their collections. Martin then transformed his bookstore into a publishing house for marketing mathematical models. Their collaboration continued after 1904 when Friedrich gained a full professorship at the TH Danzig and Martin resettled his business to Leipzig. By 1911, he issued a new seventh edition of his catalogue that advertised nearly 400 models; several of the newer ones came from his brother Friedrich.

Martin Schilling tended to follow his brother's advice in guiding the orientation of his publishing house. From him or his colleagues he received prototypes or templates for mathematical models. Thus, for example, they might send a provisional wooden frame for a thread model similar to those made earlier by Olivier. If this proved suitable, Schilling would hire a locksmith to construct appropriate metal frames. What followed afterward was for the most part in the hands of the female members of the family in the tradition of a cottage industry, as "these frames were covered with silk threads by the women of the family (mother, grandmother, aunts) in the domestic living room."[4] Thus, the "women of the family" were the invisible hands behind the scenes, an activity that was by no means marginal. For stringing silk threads, scoring lines into ready-made plaster models, or cutting out cardboard slices—work often carried out by women—required a great deal of care and precision. In terms of labor economics, model manufacturers such as Schilling (and probably Brill also) tried to ensure that as much work as possible took place in one location. Moreover, it was quite common in family publishing businesses for wives to work behind the scenes as helpers and advisors to their husbands. Likewise, their part in model making was not academic, but rather took place in a silent way. Martin Schilling, who turned his living room into a model production facility, ran a family business that drew on unpaid female employees. The economic success of his publishing house in the years between 1899 and 1920 clearly benefited from the work of these women.

Until recently, the history of science had long neglected the "silent sources" or "invisible hands" involved in scientific practices (Güttler 2013). Lately, however, a number of historians have begun to look at the other side of the story, while deconstructing tales of the genius-scientist who conceived everything on his own (Oertzen 2014; Opitz, Bergwik, and van Tiggelen 2016). In fact, many different sorts of practitioners were involved in the process of model making, though their names seldom appeared unless they happened to be professional mathematicians. Some mathematicians used models as props in staging themselves to create a special type of scientific persona for the public (Sibum 2003; Jordanova 2000). Notable examples of this among German mathematicians of the period occur in photos of Alexander Brill, Hermann Wiener, August Gutzmer, and David Hilbert. Especially remarkable is a family portrait of Hermann Wiener, surrounded by his wife and children, from about 1907 (see Figure 7.5). Here, he poses as husband, father and model maker, engaging his whole family in the process of contemplating the knowledge hidden within a mathematical model, which he holds in his hands (Sattelmacher 2021: 2).

This patriarchal portrait suggests another related theme in the history of science, namely the elevation of "head over hands," or the brilliant thinker as opposed to the mundane mechanic (note the resemblance to the portrait on the cover of Volume 4). The silence of the sources thus extends to nearly everyone who happened to be involved in actually constructing mathematical models. Thus, one often finds labels attached to models in the Brill collection that identify the "modeler," meaning the mathematician who originally designed it. Quite extensive information about some models appears in the catalogues published by the Brill and Schilling firms, which typically contain remarks such

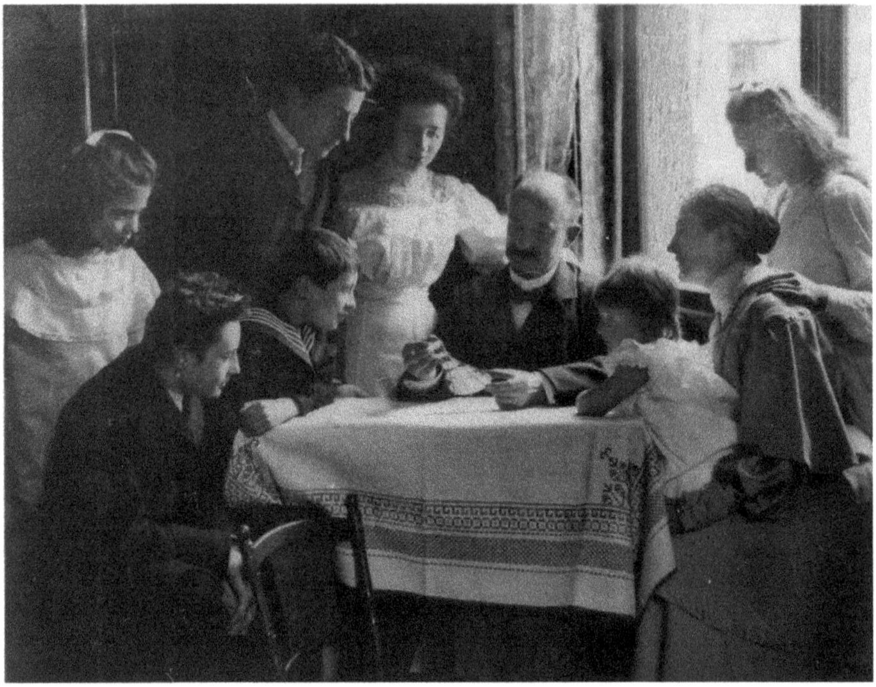

FIGURE 7.5. Hermann Gustav Wiener, surrounded by his wife and seven children. In his hands he holds a mathematical model of a rotation hyperboloid. With permission from KIT Archive, Signature 28002/517.

as made "under the supervision of A. Brill and F. Klein" or "conceived by Dr. Carl Rodenberg." Yet, never do we find any further information about the actual production process, including the reproduction and replication of models meeting high standards of aesthetic quality.

To appreciate the ways in which model makers used mathematical artifacts to represent their own standing within the discipline of mathematics, it is important to take a step backward in order to take into consideration the history of scientific collections and exhibitions. Collections of scientific instruments dating back to the early eighteenth century contributed significantly to the emergence of experimental natural sciences. By 1800, such collections began to proliferate in France alongside the spread of state-run secondary schools. In Paris, an important collection of scientific instruments gradually developed at the CNAM, which undertook the initiative of acquiring private collections containing objects useful for the courses taught there. One such collection came from Charles Muret, remembered for his large-scale relief perspectives, including one for the city of Paris and another for the (unfinished) Suez Canal. Another collection was the "Galérie Olivier," named in honor of Théodore Olivier. This included mechanisms for threading and gearing, roof trusses, stone carvings, as well as instruments for measuring, counting and drawing. This gallery also displayed a large number of Olivier's string models for paraboloids, hyperboloids, conoids, etc.

As noted earlier, in 1872 the South Kensington Museum in London acquired a large collection of Olivier thread models produced by Fabre de Lagrange. These impressive artifacts illustrating ruled surfaces found a prominent place among the geometric models on display as part of *Special Loan Collection of Scientific Apparatus*, exhibited in South Kensington from May to December 1876. As a major public event of the era, this exhibition of some 20,000 scientific artifacts attracted over a quarter million visitors. Its permanent collection, in fact, became the basis for London's present-day Science Museum. Geometry was only one of the many sections on display, but its exhibitors acquitted themselves well. Many of the exponents on display came from Germany, including Brill's cardboard models of quadric surfaces. The London Mathematical Society provided viewers with the opportunity to gaze upon Julius Plücker's exotic collection of wooden models (see Figure 0.8), restored for the occasion by Olaus Henrici, along with several others made by members of the Cambridge Modelling Club, whose custodian was James Clerk Maxwell.

The famous physicist himself contributed stereograms showing the lines of curvature of surfaces together with a stereoscope for viewing them. This type of optical equipment for visualizing models was becoming common by this time. Already in 1869, Christian Wiener studied mappings of mathematical models by means of a stereoscope for the collection in Karlsruhe. His interest at the time was mainly to keep fellow colleagues informed by sending them double-image photographs they could then view through a stereoscope. His son Hermann employed this technique often for models in his collection.

During the years they taught at the TH Munich, Brill and Klein laid the groundwork for an impressive collection of models there (see figure 7.6). This was several years before Klein began his professorial career at the University of Göttingen. Hermann Amandus Schwarz, his senior colleague, headed its mathematical seminar, which already featured a collection of models. After Schwarz left for Berlin in 1892, Klein became director of the seminar, transforming its library into a reading room ("Lesezimmer"), while gradually expanding its collection of models and instruments to include the field of geodesy. Ten years later, he oversaw its division into two separate collections, the first for models and the second for graphical instruments and calculating devices. Klein also continued another tradition due to Schwarz, who had established a modeling workshop located in a corner room in the main auditorium building of the University of Göttingen.

During the next two decades, Klein solidified his relationships abroad, beginning with a trip to the United States in 1893, during which he delivered his Evanston Colloquium Lectures (Parshall and Rowe 1994: 333–53). Five years later, Klein

FIGURE 7.6. Display cases of models at the Mathematics Institute at the Technische Hochschule in Munich, around 1912. Image courtesy of Anja Sattelmacher.

co-founded the Göttingen Association for Promotion of Applied Physics and Mathematics (Göttinger Vereinigung or GV). With the help of financing from leading industrialists and engineers, the GV effectively transformed a small university town into one of the principal hubs for scientific technology in Germany. This transformation came about soon after 1904, the year in which two prominent researchers from the Technical College Hanover accepted professorships in Göttingen: the hydrodynamics expert Ludwig Prandtl (Eckert 2019) and Carl Runge, a pioneer in numerical methods. Runge became head of the new Institute for Applied Mathematics and Mechanics and inherited the collection of mathematical instruments, whereas Klein maintained control of the models collection. He no longer had the support of Friedrich Schilling, who left for the TH Danzig in 1904; henceforth Göttingen lacked teaching expertise in descriptive geometry.

For Klein, this probably seemed a small price to pay for the expansion and modernization he now had in mind. Except for the new institute that housed Runge and Prandtl, all other mathematical activity took place in the limited facilities offered by the auditorium building. These had been adequate over the course of the first decades after it opened in 1865, but by 1900, its quarters had begun to feel increasingly cramped. Soon thereafter, however, with funding from the GV in hand, the landscape changed with new research institutes mushrooming outside the old city walls. In 1906, a new building opened its doors with facilities not only for experimental and theoretical physics but also for a new institute of applied electricity.

While overseeing this expansion, Klein began imagining an institute for mathematics located in a new building nearby and with similarly generous facilities. These would include a special room for drawing exercises, a workshop for building models, space for a larger mathematical library, plus a separate area for exhibiting the models collection. None of these features was entirely new, but the scale of these plans was highly ambitious for the time. Such a project was unusual for mathematics, since facilities like these were normally restricted to the experimental sciences. Klein's basic idea was to strengthen links between mathematics and related areas in the engineering and natural sciences. Moreover, his larger vision aimed not so much to create an independent institute but rather a mathematical center well integrated with other research fields. This he hoped to achieve through the proximity of the mathematical institute to the new institutes for physics and applied mechanics, a complex of buildings that would symbolize the modern mathematical sciences.

Financing and other arrangements for this project had advanced quite far by August 1914, when the Great War broke out. When it finally ended in November 1918, the Weimar Republic was in no position to realize the institute Klein had long imagined. When he died in June 1925, he had no inkling of any chance to revive this old plan. Not long afterward, however, his successor, Richard Courant, applied to the Rockefeller Foundation for the funding needed to finally build it. His application received approval, and in January 1930, the new institute building officially opened at the location Klein had originally planned (Siegmund-Schultze 2001). At the opening ceremony, David Hilbert lauded Courant's achievement as akin to the fairy tale of Sleeping Beauty: he was the Prince Charming who made Klein's old dream come back to life.

Courant's principal assistant, Otto Neugebauer, took over responsibility for designing the interior structure. He had long overseen the Lesezimmer, so he realized the need for a spacious new library, but he also highlighted the symbolic significance of the models by placing the collection in the upstairs foyer of the building. The actual design of the mathematical institute in Göttingen created a kind of museum for mathematical models (Sattelmacher 2014); see figure 7.7. In his description of the newly opened building, Neugebauer described the collection as "essentially Klein's work," with models and instruments placed:

> in about 50 glass cabinets, partly completely detached, partly designed as wall cabinets [. . .] The visual classification of the objects was carried out as objectively as possible, supported by an index accessible to the public, which indicates important literature relating to the models.[5]

FIGURE 7.7. The Göttingen Model Collection in 2011. Image courtesy of Anja Sattelmacher.

When he wrote this, Otto Neugebauer had only begun his career as an historian of ancient mathematics, and yet this single short quotation already reflects the approach he would later take in his historical research. Neugebauer always worked systematically to develop an inventory of artifacts that he then tried to classify "as objectively as possible" (Pyenson 1995; Rowe 2016).

The visibility of mathematical models in the institute building played a major role for Klein in its planning as well as for Neugebauer in its implementation. Whether on their way to lecture rooms, the Auditorium Maximum, or the library, students would inevitably go past the display cases filled with mathematical models. Those who stopped to look more closely would easily recognize that their contents were (and still are) systematically arranged according to various genetic relationships. The models of a particular mathematical family stood together as part of a series, making it possible to compare them. This mode of presentation was not unique to Göttingen but, in fact, Neugebauer's original arrangement has remained largely unaltered since the erection of the institute building in 1929. Already by then the once vibrant era of model making belonged to the past.

From the point of view of visual style, however, the architectural design of the Göttingen collection broke new ground. Earlier model collections, such as the one in Jena, typically employed large display cases with heavy wooden frames. Since they were generally housed in rooms with few windows, the overall atmosphere was dark and shadowy. For illumination, the showrooms were equipped with arc lamps (such as in Halle) or with an artificial skylight (in Jena). In contrast with these older collections, the Göttingen models formed an integral part of the large hallway structure at the front of the building, which featured massive windows that provided ample natural lighting even on overcast days. Moreover, the models themselves were displayed in lighter glossy showcases made mainly of glass supported by a thin metal frame. They were 2.5 meters high and mounted on walnut-veneered sockets. Since the shelves, too, were constructed in glass, the objects within the cases were visible from all sides, especially those located in the middle of the hallway. This transparency enabled a viewer not only to take in individual models while still gaining an overview sense for the full display. In this way, these geometrical objects became an integral aesthetic part of the building's architecture.

The collection of mathematical models in Göttingen was thus representative in several respects. On an overall architectural level, it already pointed forward to the use of glass and natural lighting. The smaller, but still very striking collection of models on display at the new Technical University in Munich is not only integrated into the building with transparent glass shelving, the entire building was constructed in glass with massive open spaces. At the same time, the Göttingen Mathematics Institute looked backward to an earlier tradition of the nineteenth century in which visual representation played a central role in geometry. While this changed quite suddenly with the advent of new research fields dominated by abstract concepts and more general methods, geometric models provided a link with the recent past, a symbol of the continuity of mathematical knowledge distilled in the form of special surfaces illustrating classical geometrical properties. Mathematicians no longer studied these objects per se, but they nevertheless took comfort in recognizing them as an allusion to an important historical tradition.

MOVING IMAGES AS TEACHING TOOLS

During the first decades of the twentieth century, new media emerged that played a decisive role for educating the eye via moving images. Indeed, before 1910 the inventors of cinematography thought of it primarily as an educational tool. Leading film entrepreneurs, such as Charles Pathé in France or Thomas Edison in the United States, even went so far as to claim that this recent innovation would soon replace textbooks. Even though most pedagogues regarded cinema as a passive medium, many nonetheless promoted it as a hands-on-device that could awaken "scientific curiosity." As the American film historian Oliver Gaycken has stressed, nineteenth-century popular science put a strong emphasis on engagement with the senses, including the cultivation of curiosity and wonder (Gaycken 2015: 4).

An exponent of this new direction in didactics was the Darmstadt mathematics teacher Ludwig Münch, who beginning around 1910 produced several animated films for teaching purposes. In February 1912, Münch spoke about this topic to members of a science club at his school, while showing them a series of animated drawings for topics such as the Pythagorean Theorem and the theory of conic sections. He designed his films in order to give teenagers in mathematics classes a better understanding of geometrical objects. Beginning with simpler shapes—spheres, cones, and cylinders—they would gradually gain exposure to conics, quadrics, and their mutual relationships. Münch also hoped that his animated films would eventually reach students at universities and technical colleges, especially in helping them learn to visualize basic principles of projective geometry. Münch's pedagogical films represent a continuation by new means of earlier nineteenth-century efforts to strengthen intuition through *visual instruction* (Schmidgen 2012).

After 1900, Felix Klein emerged as a strong voice calling for reforms aimed at underscoring the practical utility of mathematics from elementary school onwards. Many others joined in this reform movement, including leading mathematicians from France, Italy, and the United States. Klein identified entirely with the modernist outlook that aimed to build a strong nation with the help of a new generation of scientists and engineers. The promotion of *Anschauung* became one of the watchwords for his reform program. This went hand-in-hand with the second pillar in Klein's approach to mathematics pedagogy, what he called the "education of functional thinking." For this purpose, the visualization of functions—e.g., the movement of points on a curve in relation to time—took on an essential role. Although these goals were never entirely realized, Klein's long-term legacy as a promoter of functional thinking and a genetic approach to mathematics education remains strong even to this day (Weigand 2019).

Ludwig Münch had clearly followed these discussions, including Klein's ideas with regard to reforming mathematical teaching. In the same year he released his first films, he gave a talk entitled, "On the Education of Functional Thinking in Mathematics Lessons," in which he stressed the importance of presenting geometric forms as moving figures rather than as fixed structures. Echoing Klein's view, he pointed to geometry as the key for visualizing ("veranschaulichen") functional relations. Cinematography thus represented an ideal tool for grasping temporal relations as conveyed by moving figures. In this sense, Münch's initiative circumvented existing methods for visualizing mathematical objects, such as geometrical models.

Münch initially showed his films to fellow-mathematicians in public lectures or at conferences, where they also gained some public attention. Reports of these screenings in local newspaper articles were generally positive, but they also failed to appreciate his didactic objectives. One typically enthusiastic article called Münch's method a "significant step for the popularization of science,"[6] which was far from his intention. He hoped, of course, for more understanding from professional educators, but here, too, the reaction was disappointing. Far from sharing the enthusiasm of some newspaper reporters, some mathematicians were openly skeptical. Many doubted that viewing a film could help a young person comprehend geometrical forms; the more likely effect, they claimed, would be to induce lazy thinking. As one critic put this: "If one lets the cinematograph work instead of forcing students to think for themselves, [. . .] then this does not lead to a profound understanding of things that are in fact quite abstract."[7] These words sound very similar to general critiques of the cinema from this time. Cinematography came under attack as a "consequent product of a soul-less modernity" that did nothing to enhance imagination, but rather killed it.[8]

By this time, perceptions of models took place in a visual culture that was interested in and growing accustomed to synchronized images. The simultaneous viewing of images or projected objects—as in slide shows, stereoscopic projections, or film screenings—corresponded to the increasingly common practice of gathering a larger audience in front of an optical apparatus. By 1910, a public screening might have seemed no more unusual than sitting in a lecture hall, though an individual viewing a slide show or film clearly had a very different kind of experience than one who held a model or viewed a photograph of such an object through a stereoscope. Once these new projection techniques became available, the argument that viewing or manually handling a three-dimensional model promoted understanding of the underlying mathematics gradually lost ground. Beyond this, there were practical reasons for adopting new teaching methods. For by around 1900, the number of students at technical colleges began to increase rapidly. This factor alone surely led to an interest in different types of projection apparatus rather than making direct use of models.

Around the same time that Münch produced his films, other mathematical educators were experimenting with kinematical devices to illustrate geometrical phenomena by means of light and shadow. A notable example, patented in 1912 by Erwin Papperitz, professor for descriptive geometry at the Mining Academy in Freiberg, was an optical apparatus similar to the cinematograph. This device, however, did not project images on a flat screen but rather it was used for viewing the various cross-sections of a hyperboloid or cone in the form of a three-dimensional string model that rotated around its central axis (see figure 7.8). The technique involved the projection of images in two dimensions to make them appear as three-dimensional by using a 3D-model as the projection surface. Papperitz's intentions were similar to those of Münch, namely to develop a tool that made complex geometrical forms both visible and understandable.

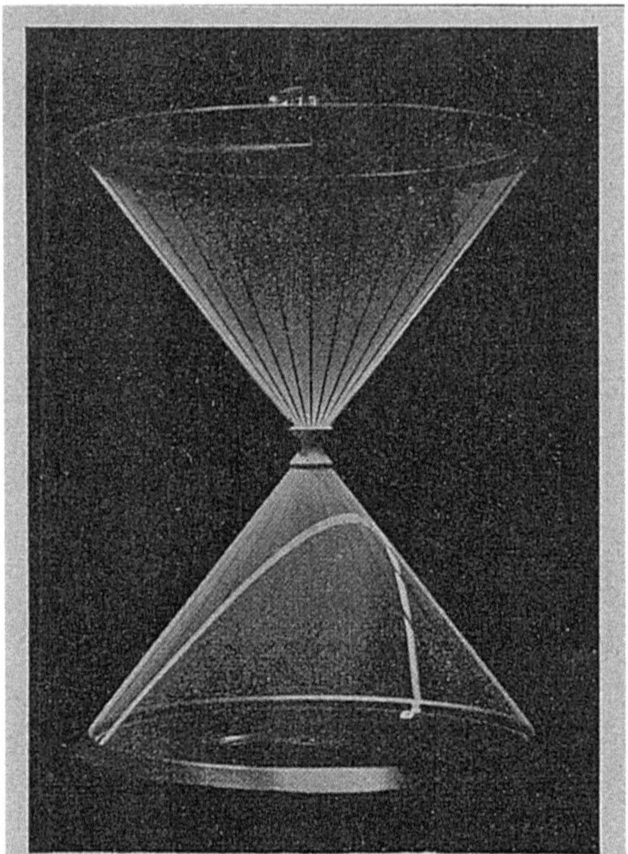

Parabel auf einem Kegel. — Parabola on a cone. — Parabole sur un cône.

FIGURE 7.8. Projected conic sections on a rotating cone produced with Papperitz's kinodiaphragmatic projection apparatus. Papperitz 1912. Image courtesy of Anja Sattelmacher.

Despite the strong claims of pioneers like Edison and Pathé regarding the potential educational benefits of cinematography, schools and institutions of higher education faced far too many practical problems that stood in the way of utilizing these types of new technologies. It thus comes as no surprise that Münch's films never became part of a teaching syllabus; nor did Papperitz's projection machine ever enter into production, despite the fact that he had signed a contract with one of Germany's leading companies in the field of educational accessories. Seen in retrospect, these inventions were ahead of their time, though cinematography would soon spawn a new entertainment industry. It also served a larger educational function in popular culture, beginning with the appearance of propaganda films made during the Great War.

CONCLUSION

During the twentieth century, the notion of a mathematical model acquired many new meanings and connotations. Most of the time such models reflected an understanding of mathematical knowledge as an abstract theory set forth in a system of axioms that simply conveyed its fundamental assumptions. This viewpoint constituted a major break with traditional axiomatics, as embodied in Euclid's *Elements*. Students raised on geometry à la Euclid conceived of axioms as self-evident truths. That ideal continued to be taught throughout the nineteenth century, but its adherents became fewer and quieter by its end. This gradual shift in perspective regarding geometrical knowledge was accompanied by a groundswell of enthusiasm for material artifacts that represented aspects of that knowledge. These were mathematical models in the mundane sense: they were not mathematical objects per se but rather images somehow related to a purely mathematical realm. These two trends—one toward abstraction, the other aiming to convey images of ideas in material form—played out over the period, even though they stood in evident opposition with one another.

From the time of Pestalozzi, direct contact with geometrical objects assumed an important role in mathematics education for children. Among leading research mathematicians, too, many spoke of the importance of cultivating an intuitional understanding of spatial relationships. The French tradition dating back to Monge lived on in the models designed by Olivier, artifacts built by Fabre de Lagrange and exhibited around the world. Leading English mathematicians, including Hirst and Arthur Cayley, carefully studied Plücker's models, while Henrici brought the then nascent tradition of German model building to Britain. The polytechnic colleges in German-speaking lands cultivated model making alongside technical drawing and projective geometry. Stepping into this open field, the Brill and Schilling families created a lucrative cottage industry that soon spread their plaster models to numerous mathematical centers. The clues left behind in these older collections of models make it possible to trace this dissemination of popular artifacts that once formed an integral part of fast-growing mathematical cultures.

Yet, the various ways of representing mathematics, especially geometry, changed over time and in accordance with conditions and context. Whereas during the early nineteenth century, building models marked a readiness to adopt an intuitive approach to geometrical objects, this tradition fell into decline by the end of the century. Over the course of this period, the techniques for representing mathematics also changed. By the end of the century, mathematical models, as modes of representation, took on a different outward appearance, becoming more mobile, transparent, lighter and larger. One notes this, for example, when viewing the models Hermann Wiener produced from 1900 onwards; these differed significantly in form and function from the plaster and cardboard models made by his cousin Alexander Brill. After 1900, mathematical models appeared in new settings rather than being merely confined to conventional display cabinets; increasingly students at universities and technical colleges saw them as two-dimensional images produced by a slide projector.

By the mid-twentieth century, very few mathematicians took any serious interest in the models in their collections, which often fell into disrepair; many were eventually lost over time. Yet, these classical artifacts from the preceding century never fully vanished, and with the advent of modern computer graphics and 3D-printing, a strong resurgence of interest in visualizing mathematical objects has taken place over the last two decades. For example, some years ago mathematicians in Göttingen began reproducing models from their collection using 3D-scanning programs and printers. These new technologies thus offer a way to breathe new life into these old artifacts, as it were, making them more relevant and accessible to contemporary mathematicians.

Nevertheless, this new wave of inventiveness does not really resolve the fundamental tension that continues to exist between a material model and the mathematical knowledge it represents. Moreover, ever since the days of Monge and Olivier one of the central motivations behind the construction of mathematical models was their presumed utility as teaching tools. Mathematicians designed and built them in most cases for didactical reasons, no matter what the level of difficulty. Many models found in the Göttingen collection, in fact, aim to convey knowledge from various realms of higher mathematics. Still, the underlying premise behind this activity claimed that these objects somehow embody mathematical ideas and thereby provide them at once with a vivid and haptic aspect. Yet, even during the heyday of model making, there appears to be little evidence that these types of objects actually assumed a central place in teaching practices. Their role throughout the course of the nineteenth century was significant, to be sure, but the history of modeling had other dimensions that went well beyond pedagogy alone. In summary, to understand the history of mathematical models during this period one must consider a variety of different factors, including the particular contexts of production, distribution, collection and exhibition, as well as the motivations of those who engaged in these practices.

NOTES

Chapter 1

1. Florence A. Yeldham, *The Teaching of Arithmetic through Four Hundred Years 1535–1935* (London: Georg G. Harran & Company 1936), 129.
2. John Brook-Smith, *Arithmetic in Theory and Practice*, 6th ed. (London: Macmillan, 1881), 255–6.
3. Ibid., 270.
4. Ibid., 148, original italics.
5. Sophia De Morgan, *Memoir of Augustus De Morgan* (London: Longmans, Green, and Co., 1882), 244.
6. Augustus De Morgan, *A Budget of Paradoxes* (London: Longmans, Green, and Co., 1872), 301.

Chapter 2

1. Isabel Maddison, "The Position of Women at Göttingen University," undated clipping, 1895, Bryn Mawr College Archives.
2. Arthur Kirchhoff, Die *akademische Frau: Gutachten hervorragender Universitätsprofessoren, Frauenlehrer und Schriftsteller über die Befähigung der Frau zum wissenschaftlichen Studium und Berufe* (Berlin: Hugo Steinitz Verlag, 1897).
3. See Philippe Nabonnand ed., *La Correspondance entre Henri Poincaré et* Gösta *Mittag-Leffler* (Basel: Birkhäuser, 1999).

Chapter 4

1. Jean Dhombres ed., *L'École normale de l'an III. Leçons de mathématiques* (Paris: Dunod, 1992), 204.

2. *Cauchy, Augustin-Louis. 1821. Cours d'Analyse de l'École royale polytechnique. Paris: Debure frères, Libraires du Roi et de la Bibliothèque du Roi.* Quotations that follow taken from this text. A modern edition is (Cauchy 1821/1990).
3. Heidegger offered the phrase, "that which is known in advance," as rendering what was originally meant by *mathema* in Greek.
4. Darwin Notebook B, DAR 121:98, available online: https://cudl.lib.cam.ac.uk/view/MS-DAR-00121/98 (accessed May 26, 2023).
5. Darwin Notebook E, DAR 124:59, available online: https://cudl.lib.cam.ac.uk/view/MS-DAR-00124/59 (accessed May 26, 2023).
6. George Boole, *Mathematical Analysis of Logic* (Cambridge: Cambridge University Press, [1847] 2009).
7. Paul du Bois-Reymond, *Die allgemeine Functionentheorie* (Tübingen: Laupp, 1882), 153.
8. "Briefe Ch. Hermite an Paul du Bois-Reymond aus den Jahren 1875–1888," *Archiv der Mathematik und Physik*, (3), 24(1916), 209.
9. Paul du Bois-Reymond, *Über die Grundlagen der Erkenntnis in den exacten Wissenschaften* (Tübingen: Laupp, 1890), sect. VIII, "*Ueber Weltanschauungen*," 121.
10. Peter Michael Harman ed., *The Scientific Letters and Papers of James Clerk Maxwell*, vols I–III (Cambridge: Cambridge University Press, 1990–2002), vol. I, 711–12.
11. Maxwell to Peter Guthrie Tait, December 11, 1867 (Harman, ed., 1990–2002: vol. II, 331).
12. Maxwell to Galton, February 26, 1879 (Harman, ed., 1990–2002: vol. III, 757).
13. Émile Boutroux, *The Contingency of the Laws of Nature*, translation by Fred Rothwell (Chicago, IL, and London: The Open Court Publishing Company, [1874] 1920), 45.
14. Émile Boutroux, *Law in Science and Philosophy*, translation by Fred Rothwell (New York: Macmillan, [1895] 1914), 77.
15. Kerry Benno, *System einer Theorie der Grenzbegriffe: Ein Beitrag zur Erkenntnistheorie* (Leipzig-Wien: Franz Deuticke, 1890), 187.
16. Ernst Mach, *Die Mechanik in ihrer Entwickelung, historisch-kritisch dargestellt* (Leipzig: Brockhaus, 1883), 437. See also Ernst Mach, *The Science of Mechanics: A Critical and Historical Exposition of Its Principles*, translation by Thomas J. McCormack (Chicago, IL: Open Court, 1893), 464.
17. Max Planck, "Die Einheit des physikalischen Weltbildes," in *Physikalische Rundblicke: Gesammelte Reden und Aufsätze* (Leipzig: Hirzel, [1908] 1922).
18. Ibid., 33.
19. Pierre Duhem, *La Théorie physique. Son objet, sa structure* (Paris: Chevalier & Rivière, 1906; reprint Paris: Vrin, 2007).
20. Ludwig Boltzmann, *Theoretical Physics and Philosophical Problems: Selected Writings*, edited by Brian McGuinness (Dordrecht and Boston, MA: Reidel, 1974), 55.
21. Henri Poincaré, *Science and Hypothesis*, translated by Mélanie Frappier, Andrea Smith, and David J. Stump (London and New York: Bloomsbury, [1902] 2018).

22. Robert Musil, "The Mathematical Man," in Burton Pike and David S. Luft eds., *Precision and Soul: Essays and Addresses* (Chicago, IL, and London: University of Chicago Press, [1913] 1990).
23. Robert Musil, *On Mach's Theories*, translation by Kevin Mulligan (München-Wien: Philosophia Verlag, [1908] 1982).
24. Ernst Mach, *Die Mechanik in ihrer Entwickelung, historisch-kritisch dargestellt* (Leipzig: Brockhaus, 1883), 431. See also Ernst Mach, *The Science of Mechanics: A Critical and Historical Exposition of Its Principles*, translation by Thomas J. McCormack (Chicago, IL: Open Court, 1893), 459.
25. Ernst Mach, *Popular Scientific Lectures*, translation by Thomas Joseph McCormack (Chicago, IL: Open Court, 1895), 188.

Chapter 5

1. Adolphe Quetelet, "Sur l'appréciation des documents statistiques, et en particulier sur l'appréciation des moyennes," *Bulletin de la Commission centrale de statistique*, 2 (1844): 205–86, on 262.
2. Gillian Beer, "'The Death of the Sun': Victorian Solar Physics and Solar Myth," in J. B. Bullen ed., *The Sun is God: Painting, Literature, and Mythology in the Nineteenth Century* (Oxford: Oxford University Press, 1989), 159–80.

Chapter 6

1. Richard Feynman, *The Feynman Lectures on Physics* (Reading, MA: Addison-Wesley Publishing Company, 1964), vol. 2, 111.
2. George Bryan, "Aeronautical Theories," *Nature*, 98 (1917): 465–7, 466.

Chapter 7

1. Fabre de Lagrange ed., *A Catalogue of a Collection of Models of Ruled Surfaces* (London: George E. Eyre and William Spottiswoode, 1872).
2. Alexander Brill to Rudolf Sturm, November 24, 1875. In "Chronicle of Alexander Brill," typographed manuscript, 1887–1935, Archive of the Technical University Munich, 41b.
3. August Gutzmer, 1904: *Geschichte der Deutschen Mathematiker-Vereinigung von ihrer Begründung bis zur Gegenwart* (Leipzig: Teubner, 1904), 20–1.
4. Transcript of an interview of Karin Richter with Juliane Vogel, descendant of Friedrich Schilling in Halle, 2008, 1.
5. Otto Neugebauer, "Das mathematische Institut der Universität Göttingen," *Die Naturwissenschaften*, 18, 1 (1930): 2.
6. Ludwig Münch, "Der Kinematograph im mathematischen Unterricht," *Kommunale Praxis*, 11 (1912): 329.
7. Karl Stockmeyer, "Gehört der Kinematograph in den Mathematikunterricht?" in *Südwestdeutsche Schulblätter*, 30, 6/7 (1913): 450–6, 454.
8. Cf. Franz Pfempfert, "Der Kinematograph als Erzieher," in Jörg Schweinitz ed., *Prolog vor dem Film: Nachdenken über ein neues Medium, 1909–1914* (Leipzig: Reclam, 1992), 165–9.

BIBLIOGRAPHY

Albers, Donald J., Gerald L. Alexanderson, and Constance Reid, eds. (1987), *International Mathematical Congresses: An Illustrated History, 1893–1986*, New York: Springer.

Alder, Ken. (1999), "French Engineers Before Professionals; or, How Meritocracy Made Knowledge Objective," in William Clark, Jan Golinski, and Simon Schaffer eds., *The Sciences in Enlightened Europe*, 94–125, Chicago, IL: University of Chicago Press.

Anderson, Nancy A., and Michael R. Dietrich, eds. (2012), *The Educated Eye: Visual Culture and Pedagogy in the Life Sciences*, Hanover, NH: Dartmouth College Press.

Barbin, Évelyne, Marta Menghini, and Klaus Volkert, eds. (2019), *Descriptive Geometry: The Spread of a Polytechnic Art*, Heidelberg: Springer.

Barrow-Green, June (2021), "'Knowledge Gained by Experience': Olaus Henrici—Engineer, Geometer and Maker of Mathematical Models," *Historia Mathematica*, 54: 71–6.

Begehr, Heinrich, Helmut Koch, Jürg Kramer, Norbert Schappacher, and Ernst-Jochen Thiele, eds. (1998), *Mathematics in Berlin*, Heidelberg: Springer.

Belhoste, Bruno (1991), *Augustin-Louis Cauchy: A Biography*, New York: Springer.

Bellone, Enrico (1980), *A World of Paper: Studies on the Second Scientific Revolution*, Cambridge, MA: MIT Press.

Bienvenu, Richard T., and Mordechai Feingold, eds. (1991) *In the Presence of the Past: Essays in Honor of Frank Manuel*, Dordrecht: Kluwer Academic.

Bierende, Edgar, and Frank Loose (2018), "Abgüsse in Ausstellungen: Brills Modelle in München und Chicago," in Ernst Seidl, Frank Loose, and Edgar Bierende eds., *Mathematik mit Modellen: Alexander von Brill und die Tübinger Modellsammlung*, 91–102, Tübingen: Museum der Universität Tübingen.

Blackmore, John, ed. (1992), *Ernst Mach—A Deeper Look: Documents and New Perspectives*, Dordrecht: Kluwer-Springer.

Blackmore, John, ed. (1995), *Ludwig Boltzmann: His Later Life and Philosophy, 1900–1908, Book One: A Documentary History*, Dordrecht: Kluwer-Springer.

Blondel, Christine (1982), *Ampère et la création de l'électrodynamique*, Paris: Bibliothèque Nationale.

Bloor, David (1981), "Hamilton and Peacock on the Essence of Algebra," in Herbert Mehrtens, Henk Bos, and Ivo Schneider eds., *Social History of Nineteenth-Century Mathematics*, 202–32, Boston, MA: Birkhäuser-Springer.

Bottazzini, Umberto (1986), *The Higher Calculus: A History of Real and Complex Analysis from Euler to Weierstrass*, Heidelberg: Springer.

Bottazzini, Umberto, and Amy Dahan-Dalmedico, eds. (2001), *Changing Images in Mathematics from the French Revolution to the New Millennium*, London: Routledge.

Brechenmacher, Frédéric (2022), "Knowing by Drawing: Geometric Material Models in Nineteenth Century France," in Michael Friedman and Karin Krauthausen eds., *Model and Mathematics: From the 19th to the 21st Century*, 53–143, Cham: Birkhäuser.

Brenni, Pablo (2010), "The Evolution of Teaching Instruments and Their Use between 1800 and 1930," *Science and Education*, 21 (2): 191–226.

Brush, Stephen. G. (1983), *Statistical Physics and the Atomic Theory of Matter, from Boyle and Newton to Landau and Onsager*, Princeton, NJ: Princeton University Press.

Brush, Stephen G. (1987), "The Nebular Hypothesis and the Evolutionary Worldview," *History of Science*, 25 (3): 245–78.

Buchwald, Jed Z. (1989), *The Rise of the Wave Theory of Light: Optical Theory and Experiment in the Early Nineteenth Century*, Chicago, IL: University of Chicago Press.

Calinger, Robert, ed. (1996), *Vita Mathematica: Historical Research and Integration with Teaching*, Providence, RI: Mathematical Association of America.

Caneva, Kenneth (1980), "Ampère, the Etherians, and the Oersted Connexion," *British Journal for the History of Science*, 13 (2): 121–38.

Caneva, Kenneth (1997), "Physics and Naturphilosophie: A Reconnaissance," *History of Science*, 35 (1): 35–106.

Canguilhem, Georges ([1965] 2008), *Knowledge of Life*, ed. Paola Marrati and Todd Meyers, trans. Stefanos Geroulanos and Daniela Ginsburg, New York: Fordham University Press.

Carus, André W. (2007), *Carnap and Twentieth-Century Thought: Explication and Enlightenment*, Cambridge: Cambridge University Press.

Cassirer, Ernst ([1910] 2003), *Substance and Function*, trans. William Curtis Swabey and Marie Collins Swabey, Mineola, NY: Dover.

Cassirer, Ernst (1956), *Determinism and Indeterminism in Modern Physics: Historical and Systematic Studies of the Problem of Causality*, trans. O. Theodor Benfey, New Haven, CT: Yale University Press.

Cauchy, Augustin-Louis (1821/1990), *Cours d'analyse de l'Ecole Royale Polytechnique, Première Parie, Analyse Algébrique*, Umberto Bottazzini ed., Bologna: Editions CLUEB.

Clark, William, Jan Golinski, and Simon Schaffer, eds. (1999), *The Sciences in Enlightened Europe*, Chicago, IL: University of Chicago Press.

Cline Cohen, Patricia (1999), *A Calculating People: The Spread of Numeracy in Early America*, New York: Routledge.

Cooke, Roger (1984), *The Mathematics of Sonya Kovalevskaya*, New York: Springer.

Corry, Leo (2004a), *David Hilbert and the Axiomatization of Physics (1898–1918): From* Grundlagen der Geometrie *to* Grundlagen der Physik, Archimedes: New Studies in the History and Philosophy of Science and Technology, #10, Dordrecht: Kluwer Academic.

Corry, Leo (2004b), *Modern Algebra and the Rise of Mathematical Structures*, 2nd edn, Basel: Birkhäuser.
Crosland, Maurice (1967), *The Society of Arcueil: A View of French Science at the Time of Napoleon I*, London: Heinemann.
Darrigol, Olivier (2000), *Electrodynamics from Ampère to Einstein*, Oxford: Oxford University Press.
Darrigol, Olivier (2012), *A History of Optics from Greek Antiquity to the Nineteenth Century*, Oxford: Oxford University Press.
Daston, Lorraine (1986), "The Physicalist Tradition in Early Nineteenth-Century French Physics," *Studies in History and Philosophy of Science*, 17 (3): 269–95.
Dauben, Joseph W. (1979), *Georg Cantor: His Mathematics and Philosophy of the Infinite*, Princeton, NJ: Princeton University Press.
Dauben, Joseph W. (2002), "Internationalizing Mathematics East and West: Individuals and Institutions in the Emergence of a Modern Mathematical Community in China," in Karen Hunger Parshall and Adrian C. Rice eds., *Mathematics Unbound: The Evolution of an International Mathematical Research Community, 1800–1945*, 253–85, History of Mathematics, #23, Providence, RI/London: American Mathematical Society/London Mathematical Society.
de Chadarevian, Soraya, and Nick Hopwood, eds. (2004), *Models: The Third Dimension of Science*, Stanford, CA: Stanford University Press.
de Courtenay, Nadine (2018), "Boltzmann et le réalisme scientifique (I): de l'épistémologie des sciences de la nature à la philosophie comme clarification du langage," *Philosophie*, 137: 18–38.
Denniss, John (2009), "Learning Arithmetic: Textbooks and Their Users in England 1500–1900," in Eleanor Robson and Jaqueline Stedall eds., *The Oxford Handbook of the History of Mathematics*, 448–67, Oxford: Oxford University Press.
Eckert, Michael (2013), *Arnold Sommerfeld: Science, Life and Turbulent Times 1868–1951*, Heidelberg: Springer.
Eckert, Michael (2019), *Ludwig Prandtl: A Life for Fluid Mechanics and Aeronautical Research*, Heidelberg: Springer.
Enros, Philip (1981), "Cambridge University and the Adoption of Analytics in Early Nineteenth-Century England," in Herbert Mehrtens, Henk Bos, and Ivo Schneider eds., *Social History of Nineteenth-Century Mathematics*, 135–48, Boston, MA: Birkhäuser.
Enros, Philip (1983), "The Analytical Society (1812–1813): Precursor of the Renewal of Cambridge Mathematics," *Historia Mathematica*, 10: 24–47.
Epple, Moritz (1999), *Die Entstehung der Knotentheorie: Kontexte und Konstruktionen einer modernen mathematischen Theorie*, Braunschweig: Vieweg.
Epple, Moritz (2006), "Felix Hausdorff's Considered Empiricism," in José Ferreirós and Jeremy J. Gray eds., *The Architecture of Modern Mathematics: Essays in History and Philosophy*, 263–90, Oxford: Oxford University Press.
Feichtinger, Johannes, Franz L. Fillafer, and Jan Surman, eds. (2018), *The Worlds of Positivism: A Global Intellectual History, 1770–1930*, Cham: Palgrave Macmillan.
Ferreirós, José ([1999] 2007), *Labyrinth of Thought: A History of Set Theory and Its Role in Modern Mathematics*, 2nd edn, Basel: Birkhäuser.
Ferreirós, José (2016), *Mathematical Knowledge and the Interplay of Practices*, Princeton, NJ: Princeton University Press.
Ferreirós, José, and Jeremy J. Gray, eds. (2006), *The Architecture of Modern Mathematics: Essays in History and Philosophy*, Oxford: Oxford University Press.

Flood, Raymond, Mark McCartney, and Andrew Whitaker (2014), *James Clerk Maxwell: Perspectives on His Life and Work*, Oxford: Oxford University Press.
Flood, Raymond, Adrian Rice, and Robin Wilson, eds. (2011), *Mathematics in Victorian Britain*, Oxford: Oxford University Press.
Foucault, Michel ([1966] 2002), *The Order of Things: An Archaeology of the Human Sciences*, London: Routledge.
Foucault, Michel ([1975] 2012), *Discipline and Punishment: The Birth of the Prison*, trans, Alan Sheridan, New York: Vintage Books.
Foucault, Michel (1984), "What is Enlightenment?," in Paul Rabinow ed., *The Foucault Reader*, 32–50, New York: Pantheon Books.
Fox, Robert (1974), "The Rise and Fall of Laplacian Physics," *Historical Studies in the Physical Sciences*, 4: 89–136.
Fox, Robert (2012), *The Savant and the State: Science and Cultural Politics in Nineteenth-Century France*, Baltimore, MD: Johns Hopkins University Press.
Friedman, Michael (2018), *A History of Folding in Mathematics: Mathematizing the Margins*, Berlin: Springer.
Friedman, Michael, and Karin Krauthausen, eds. (2022), *Model and Mathematics: From the 19th to the 21st Century*, Cham: Birkhäuser.
Friedman, Michael, and Alfred Nordmann, eds. (2006), *The Kantian Legacy in Nineteenth-Century Science*, Cambridge, MA: MIT Press.
Gaycken, Oliver (2015), *Devices of Curiosity*, Oxford: Oxford University Press.
Gigerenzer, Gerd, Zeno Swijtink, Theodore Porter, Lorraine Daston, John Beatty, and Lorenz Krüger, eds. (1989), *The Empire of Chance: How Probability Changed Science and Everyday Life*, Cambridge: Cambridge University Press.
Gillies, Donald, ed. (1992), *Revolutions in Mathematics*, Oxford: Clarendon Press.
Goldstein, Catherine, Nobert Schappacher, and Joachim Schwermer, eds. (2007), *The Shaping of Arithmetic after C.F. Gauss's Disquisitiones Arithmeticae*, Heidelberg: Springer.
Gordon, John Steele (2002), *A Thread across the Ocean: The Heroic Story of the Transatlantic Cable*, New York: Simon & Schuster.
Grabiner, Judith V. (1981), *The Origins of Cauchy's Rigorous Calculus*, Mineola, NY: Dover.
Grattan-Guinness, Ivor (1984), "Work for the Workers: Advances in Engineering Mechanics and Instruction in France, 1800–1830," *Annals of Science*, 41 (1): 1–33.
Grattan-Guinness, Ivor (1990), *Convolutions in French Mathematics, 1800–1840*, 3 vols, Basel: Birkhäuser.
Gray, Jeremy (1992), "The Nineteenth-Century Revolution on Mathematical Ontology," in Donald Gillies ed., *Revolutions in Mathematics*, 226–48, Oxford: Clarendon Press.
Gray, Jeremy (2000), *The Hilbert Challenge*, Oxford: Oxford University Press.
Gray, Jeremy (2008), *Plato's Ghost: The Modernist Transformation of Mathematics*, Princeton, NJ: Princeton University Press.
Gray, Jeremy (2011), *Worlds out of Nothing: A Course on the History of Geometry in the 19th Century*, 2nd edn, Heidelberg: Springer.
Gray, Jeremy (2013), *Henri Poincaré: A Scientific Biography*, Princeton, NJ: Princeton University Press.
Gray, Jeremy (2018), *A History of Abstract Algebra: From Algebraic Equations to Modern Algebra*, Heidelberg: Springer.
Gray, Jeremy (2021), *Change and Variations: A History of Differential Equations to 1900*, Heidelberg: Springer.

Güttler, Nils (2013), "Unsichtbare Hände," *Archiv für die Geschichte des Buchwesens* 68: 113–33.

Hacking, Ian (1983), *The Taming of Chance*, Cambridge: Cambridge University Press.

Harman, Peter Michael (1998), *The Natural Philosophy of James Clerk Maxwell*, Cambridge: Cambridge University Press.

Harris, Mary (1997), *Common Threads: Women, Mathematics and Work*, Stoke on Trent: Trentham Books.

Heidegger, Martin ([1938] 2002), "The Age of the World Picture," in *Off the Beaten Track*, ed. and trans. Julian Young and Kenneth Haynes, 57–85, Cambridge: Cambridge University Press.

Henderson, Linda Dalrymple ([1983] 2014), *The Fourth Dimension and Non-Euclidean Geometry in Modern Art*, Princeton, NJ: Princeton University Press; 2nd enlarged edition, Cambridge, MA: MIT Press.

Hodgkin, Luke (1981), "Mathematics and Revolution from Lacroix to Cauchy", in Herbert Mehrtens, Henk Bos, and Ivo Schneider eds., *Social History of Nineteenth-Century Mathematics*, 50–71, Boston, MA: Birkhäuser-Springer.

Hollings, Christopher, Ursula Martin, and Adrian Rice (2018), *Ada Lovelace: The Making of a Computer Scientist*, Oxford: Bodleian Library.

Hoppit, Julian (1993), "Reforming Britain's Weights and Measures, 1660–1824," *The English Historical Review*, 108 (426): 82–104.

Horkheimer, Max, and Theodor W. Adorno ([1969] 2002), *The Dialectic of the Enlightenment: Philosophical Fragments*, ed. Gunzelin Schmid Noerr and trans. Edmund Jephcott, Stanford, CA: Stanford University Press.

Howson, A. Geoffrey (1982), *A History of Mathematics Education in England*, Cambridge: Cambridge University Press.

Hunt, Bruce J. (2021), *Imperial Science: Cable Telegraphy and Electrical Physics in the Victorian British Empire*, Cambridge: Cambridge University Press.

Jahnke, Hans Niels, ed. (2003), *A History of Analysis*, History of Mathematics, #24, Providence, RI/London: American Mathematical Society/London Mathematical Society.

Jahnke, H. N., and M. Otte, eds. (1981), *Epistemological and Social Problems of the Sciences in the Early Nineteenth Century*, Dordrecht: Reidel.

Jardine, Nicholas (1999), "Inner History; or, How to End Enlightenment," in William Clark, Jan Golinski, and Simon Schaffer eds., *The Sciences in Enlightened Europe*, 477–94, Chicago, IL: University of Chicago Press.

Jones, Alexander, Christine Proust, and John M. Steele, eds. (2016), *A Mathematician's Journeys: Otto Neugebauer and Modern Transformations of Ancient Science*, New York: Springer.

Jones, Claire G., Alison E. Martin, and Alexis Wolf, eds. (2022), *The Palgrave Handbook of Women and Science since 1660*, Cham: Springer Nature.

Jordanova, Ludmilla (2000), *Defining Features: Scientific and Medical Portraits, 1660–2000*, London: Reaktion Books.

Jungnickel, Christa, and Russell McCormmach (2016), *The Second Physicist: On the History of Theoretical Physics in Germany*, New York: Springer.

Kidwell, Peggy A. (1996), "American Mathematics Viewed Objectively: The Case of Geometric Models," in Robert Calinger ed., *Vita Mathematica: Historical Research and Integration with Teaching*, 197–207, Providence, RI: Mathematical Association of America.

Kidwell, Peggy A., Amy Ackerberg-Hastings, and David L. Roberts (2008), *Tools of American Mathematics Teaching, 1800–2000*, Washington, DC: Smithsonian Institution Press.

Klein, Ursula (2012), "Artisanal-Scientific Experts in Eighteenth-Century France and Germany," *Annals of Science*, 69 (3): 303–6.

Koblitz, Ann Hibner (1983), A *Convergence of Lives: Sofia Kovalevskaia: Scientist, Writer, Revolutionary*, Boston, MA: Birkhäuser.

Koselleck, Reinhart ([1979] 2004), *Futures Past: On the Semantics of Historical Time*, trans. Keith Tribe, New York: Columbia University Press.

Kula, Witold (1986), *Measures and Men*, trans. R. Szreter, Princeton, NJ: Princeton University Press.

Lehto, Olli (1998), *Mathematics without Borders: A History of the International Mathematical Union*, New York: Springer.

Lützen, Jesper (2003), "The Foundation of Analysis in the 19th Century," in Hans Niels Jahnke ed., *A History of Analysis*, 155–95, History of Mathematics, #24, Providence, RI/London: American Mathematical Society/London Mathematical Society.

Mehrtens, Herbert (1990), *Moderne–Sprache–Mathematik*, Frankfurt-am-Main: Suhrkamp.

Mehrtens, Herbert (2004), "Mathematical Models," in Soraya de Chadarevian and Nick Hopwood eds., *Models: The Third Dimension of Science*, 276–306, Stanford, CA: Stanford University Press.

Mehrtens, Herbert, Henk Bos, and Ivo Schneider, eds. (1981), *Social History of Nineteenth-Century Mathematics*, Boston, MA: Birkhäuser-Springer.

Merzbach, Uta (2018), *Dirichlet: A Mathematical Biography*, Basel: Birkhäuser.

Moktefi, Amirouche (2011), "*Geometry: The Euclid Debate*," in Raymond Flood, Adrian Rice, and Robin Wilson eds., *Mathematics in Victorian Britain*, 320–36, Oxford: Oxford University Press.

Moore, Gregory H. (1982), *Zermelo's Axiom of Choice: Its Origins, Development and Influence*, New York: Springer.

Moore, James, ed. (1989), *History, Humanity and Evolution: Essays for John C. Greene*, Cambridge: Cambridge University Press.

North, John D. (2008), *Cosmos: An Illustrated History of Astronomy and Cosmology*, Chicago, IL: University of Chicago Press.

Nye, Mary Jo. (1979), "The Boutroux Circle and Poincaré's Conventionalism," *Journal of the History of Ideas*, 40 (1): 107–20.

Nye, Mary Jo, ed. (2003), *The Cambridge History of Science, vol. 5: Modern Physical and Mathematical Sciences*, Cambridge: Cambridge University Press.

Oertzen, Christine von (2014), *Science, Gender, and Internationalism*, New York: Palgrave Macmillan.

Opitz, Donald L., Staffan Bergwik, and Brigitte van Tiggelen (2016), *Domesticity in the Making of Modern Science*, Basingstoke: Palgrave Macmillan.

Parshall, Karen Hunger (2006), *James Joseph Sylvester: Jewish Mathematician in a Victorian World*, Baltimore, MD: Johns Hopkins University Press.

Parshall, Karen Hunger, and Adrian C. Rice, eds. (2002), *Mathematics Unbound: The Evolution of an International Mathematical Research Community, 1800–1945*, History of Mathematics, #23, Providence, RI/London: American Mathematical Society/London Mathematical Society.

Parshall, Karen Hunger, and David E. Rowe (1994), *The Emergence of the American Mathematical Research Community, 1876–1900. J.J. Sylvester, Felix Klein, and E.H. Moore*, History of Mathematics, #8, Providence, RI/London: American Mathematical Society/London Mathematical Society.

Petri, Birgit, and Norbert Schappacher (2007), "On Arithmetization," in Catherine Goldstein, Norbert Schappacher, and Joachim Schwermer eds., *The Shaping of*

Arithmetic after C. F. Gauss's Disquisitiones Arithmeticae, 343–74, Heidelberg: Springer.
Porter, Theodore M. (1986), *The Rise of Statistical Thinking, 1820–1900*, Princeton, NJ: Princeton University Press.
Preston, John, ed. (2021), *Interpreting Mach: Critical Essays*, Cambridge: Cambridge University Press.
Pyenson, Lewis (1995), "Inventory as a Route to Understanding: Sarton, Neugebauer, and Sources," *History of Science*, 33 (3): 253–82.
Ramanujam, R., and K. Subramaniam, eds. (2012), *Mathematics Education in India: Status and Outlook*, Mumbai: Homi Bhaba Centre for Science Education.
Rheinberger, Hans-Jörg (2010), *On Historicizing Epistemology: An Essay*, trans. David Fernbach, Palo Alto, CA: Stanford University Press.
Richards, Joan L. (1988), *Mathematical Visions: The Pursuit of Geometry in Victorian England*, San Diego, CA: Academic Press.
Richards, Joan L. (1991), "Rigor and Clarity: Foundations of Mathematics in France and England, 1800–1840," *Science in Context*, 4 (2): 297–319.
Richards, Joan L. (2006), "Historical Mathematics in the French Eighteenth Century," *Isis*, 97 (4): 700–13.
Robson, Eleanor, and Jaqueline Stedall, eds. (2009), *The Oxford Handbook of the History of Mathematics*, Oxford: Oxford University Press.
Rowe, David E. (2003), "Mathematical Schools, Communities, and Networks," in Mary Jo Nye ed., *The Cambridge History of Science, vol. 5: Modern Physical and Mathematical Sciences*, 113–32, Cambridge: Cambridge University Press.
Rowe, David E. (2013), "Mathematical Models as Artefacts for Research: Felix Klein and the Case of Kummer Surfaces," *Mathematische Semesterberichte*, 60: 1–24.
Rowe, David E. (2016), "From Graz to Gottingen: Neugebauer's Early Intellectual Journey," in Alexander Jones, Christine Proust, and John M. Steele eds., *A Mathematician's Journeys: Otto Neugebauer and Modern Transformations of Ancient Science*, 1–59, New York: Springer.
Rowe, David E. (2018), *A Richer Picture of Mathematics: The Göttingen Tradition and Beyond*, New York: Springer.
Rowe, David E. (2022), "Models from the Nineteenth Century Used for Visualizing Optical Phenomena and Line Geometry," in Michael Friedman and Karin Krauthausen eds., *Model and Mathematics: From the 19th to the 21st Century*, 177–201, Cham: Birkhäuser.
Rowe, David E., and Mechthild Koreuber (2020), *Proving it Her Way: Emmy Noether, A Life in Mathematics*, Heidelberg: Springer.
Royle, Tony (2020), *The Flying Mathematicians of World War I*, Montreal: McGill-Queen's University Press.
Russ, Steve (2004), *The Mathematical Works of Bernard Bolzano*, Oxford: Oxford University Press.
Sasaki, Chikara (2002), "The Emergence of the Japanese Mathematical Community in the Modern Western Style, 1855–1945," in Karen H. Parshall and Adrian C. Rice eds., *Mathematics Unbound: The Evolution of an International Mathematical Research Community, 1800–1945*, 229–52, History of Mathematics, #23, Providence, RI/ London: American Mathematical Society/London Mathematical Society.
Sattelmacher, Anja (2014), "Zwischen Ästhetisierung und Historisierung," *Mathematische Semesterberichte*, 61 (2): 131–43.
Sattelmacher, Anja (2019), "Männer mit Modellen. Porträts von Mathematikern um 1900," in Christian Vogel and Sonja E. Nökel eds., *Gesichter der Wissenschaft.*

Repräsentanz und Performanz von Gelehrten in Porträts, 115–31, Göttingen: Wallstein.
Sattelmacher, Anja (2021), *Anschauen, Anfassen, Auffassen. Eine Wissensgeschichte mathematischer Modelle*, Wiesbaden: Springer.
Scaife, W. Garrett (2000), *From Galaxies to Turbines: Science, Technology and the Parsons Family*, London: Routledge.
Schaffer, Simon (1989), "The Nebular Hypothesis and the Science of Progress," in James Moore ed., *History, Humanity and Evolution: Essays for John C. Greene*, 131–64, Cambridge: Cambridge University Press.
Schaffer, Simon (1994), "Babbage's Intelligence: Calculating Engines and the Factory System," *Critical Inquiry*, 21 (1): 203–27.
Schappacher, Norbert (2022), *Framing Global Mathematics: The International Mathematical Union between Theorems and Politics*, Cham: Springer Nature.
Schmidgen, Henning (2012), "Cinematography without Film," in Nancy A. Anderson and Michael R. Dietrich eds., *The Educated Eye: Visual Culture and Pedagogy in the Life Sciences*, 94–121, Hanover, NH: Dartmouth College Press.
Scholz, Erhard, ed. (2001), *Hermann Weyl's* Raum-Zeit-Materie *and a General Introduction to His Scientific Work*, Basel: Birkhaeuser.
Schubring, Gert (2005), *Conflicts between Generalization, Rigor, and Intuition*, Heidelberg: Springer.
Schweber, Silvan S. (1991), "Auguste Comte and the Nebular Hypothesis," in Richard T. Bienvenu and Mordechai Feingold eds., *In the Presence of the Past: Essays in Honor of Frank Manuel*, 131–91, Dordrecht: Kluwer Academic.
Seidl, Ernst, Frank Loose, and Edgar Bierende, eds. (2018), *Mathematik mit Modellen: Alexander von Brill und die Tübinger Modellsammlung*, Tübingen: Museum der Universität Tübingen.
Senthil Babu, D. (2012), "Indigenous Traditions and the Colonial Encounter: A Historical Perspective on Mathematics Education in India," in R. Ramanujam and K. Subramaniam eds., *Mathematics Education in India: Status and Outlook*, 37–62, Mumbai: Homi Bhaba Centre for Science Education.
Sibum, Otto H. (2003), "Experimentalists in the Republic of Letters," *Science in Context*, 16 (1/2): 89–120.
Siegmund-Schultze, Reinhard (2001), *Rockefeller and the Internationalization of Mathematics between the Two World Wars*, Basel: Birkhäuser.
Small, Robin (2017), *Nietzsche in Context*, 2nd edn, London: Routledge.
Smith, Crosbie, and M. Norton Wise (1989), *Energy & Empire: A Biographical Study of Lord Kelvin*, Cambridge: Cambridge University Press.
Stanley, Matthew (2015), *Huxley's Church and Maxwell's Demon: From Theistic Science to Naturalistic Science*, Chicago, IL: University of Chicago Press.
Steinle, Friedrich (2016), *Exploratory Experiments: Ampère, Faraday, and the Origins of Electrodynamics*, Pittsburgh, PA: University of Pittsburgh Press.
Stigler, Stephen M. (1986), *The History of Statistics: The Measurement of Uncertainty before 1900*, Cambridge MA: Harvard University Press.
Stigler, Stephen M. (1999), *Statistics on the Table: The History of Statistical Concepts and Methods*, Cambridge, MA: Harvard University Press.
Swade, Doran (2000), *The Cogwheel Brain: Charles Babbage and the Quest to Build the First Computer*, Boston, MA: Little, Brown.
te Heesen, Anke (2002), *The World in a Box. The Story of an Eighteenth-Century Picture Encyclopaedia*, Chicago: University of Chicago Press.

Tobies, Renate (2021), *Felix Klein: Visions for Mathematics, Applications, and Education*, Heidelberg: Springer.
Tobies, Renate (2022), "Internationalism and Women Mathematicians at the University of Göttingen," in Claire G. Jones, Alison E. Martin, and Alexis Wolf eds., *The Palgrave Handbook of Women and Science since 1660*, 223–43, Cham: Springer Nature.
Tresch, John (2012), *The Romantic Machine: Utopian Science and Technology after Napoleon*, Chicago, IL: University of Chicago Press.
Turner, Steven (1981), "The Prussian Professoriate and the Research Imperative, 1790–1840," in H. N. Jahnke and M. Otte eds., *Epistemological and Social Problems of the Sciences in the Early Nineteenth Century*, 109–21, Dordrecht: Reidel.
Uebel, Thomas (2021), "Ernst Mach's Enlightenment Pragmatism: History and Economy in Scientific Cognition," in John Preston ed., *Interpreting Mach: Critical Essays*, 84–102, Cambridge: Cambridge University Press.
Ueno, Kenji (2012), "Mathematics Teaching Before and After the Meiji Restoration," *ZDM Mathematics Education*, 44 (4): 473–81.
van Dalen, Dirk (2013), *L.E.J. Brouwer——Topologist, Intuitionist, Philosopher: How Mathematics is Rooted in Life*, London: Springer.
Vogel, Christian, and Sonja E. Nökel, eds. (2019), *Gesichter der Wissenschaft. Repräsentanz und Performanz von Gelehrten in Porträts*, Göttingen: Wallstein.
Volkert, Klaus (2022), "Wilhelm Fiedler and His Models:– The Polytechnic Side," in Michael Friedman and Karin Krauthausen eds., *Model and Mathematics: From the 19th to the 21st Century*, 145–75, Cham: Birkhäuser.
Weigand, Hans-Georg, William McCallum, Marta Menghini, Michael Neubrand, and Gert Schubring, eds. (2019), *The Legacy of Felix Klein*, Cham: Springer Nature.
Whitmer, Kelly Joan (2015), *The Halle Orphanage as Scientific Community Observation, Eclecticism, and Pietism in the Early Enlightenment*, Chicago, IL: University of Chicago Press.
Wise, M. Norton, and Crosbie Smith (1989–90), "Work and Waste: Political Economy and Natural Philosophy in Nineteenth-Century Britain I–III," *History of Science*, 27 (3): 263–301; 27 (4): 391–449; 28 (3): 221–61.
Woolmar, Christian (2008), *Fire & Steam: A New History of the Railways in Britain*, London: Atlantic Books.
Wussing, Hans ([1968] 1984), *The Genesis of the Abstract Group Concept*, trans. Abe Shenitzer, Cambridge, MA: MIT Press.
Yandell, Benjamin (2002), *The Honors Class: Hilbert's Problems and their Solvers*, Wellesley, MA: A. K. Peters.

INDEX

Locators in *italics* refer to illustrations. Additional information may be found under named entries